Arduino Nano
VLF Metal Detector
Project

George Overton

An Inner Workings Publication

The very-low-frequency (VLF) metal detector presented in this book is an open-source design. This includes both the hardware and software. It is intended to be constructed for personal use, and not to be distributed as a commercial product without permission from the author.

Some studies have indicated that devices which emit radio frequencies may be linked to certain health problems, including disruption of pacemakers. Although there has been no indication that metal detectors are among these devices, health safety is the responsibility of the user.

Neither the author nor the publisher of this book shall be held responsible, whether legal, financial, or any other way, for any damages incurred by the use of the contents of this book.

This book was put together using LibreOffice Writer, with body text in Times New Roman 10pt, URW Chancery L (various sizes) for titles, DejaVu Sans for software, lists and tables, and Flipside BRK for the book title. The equations were created using the TexMaths LaTeX extension in LibreOffice.

Line art was drawn with Inkscape, and images manipulated with GIMP.

Circuit diagrams, PCB layout, Gerber and NC drill files were generated with KiCad 6.0.

Embedded software was created with the Arduino IDE, and uploaded to the processor using a pre-programmed bootloader in the Arduino Nano via a Mini-B USB cable.

3D printed parts were initially created with FreeCAD, then stl files were sliced in Repetier-Host using CuraEngine, and finally printed with a GEEETech i3 Pro B.

SPICE-based circuit simulations were performed using LTspice.

Linux Mint was the only operating system used during the development of the Arduino Nano VLF Metal Detector Project, and to create the contents of this book.

Table of Contents

"A long time ago, people who sacrificed their sleep, family, food, laughter and other joys of life were called saints. Now they are called engineers."

                                                                    --- Engineering fact

List of Figures

List of Tables

"Any circuit design must contain at least one part which is obsolete, two parts which are unobtainable, and three parts which are still under development."

--- An inescapable consequence of Murphy's Law

About this Book

This book is intended for Arduino users who have already mastered the basics of programming, and for those who have at least an elementary knowledge of electronics, and a particular interest in metal detecting technology.

It is assumed that the reader has progressed beyond the level of flashing LEDs and generally testing various random projects designed to show the capabilities of the Arduino platform, and is now ready to construct something more advanced that will be of real practical use.

The project presented here is for a very-low-frequency (VLF) induction balance (IB) metal detector with a professional level of performance. The majority of Arduino-based metal detector designs that can be found on the Internet are of the pulse-induction (PI) type, although there are a few relatively simplistic VLF circuits with a fairly low level of performance. The aim of this particular project is to provide a design that is flexible enough to make use of a large range of commercial coils, and to provide target identification on a number of different displays. The aim is *flexibility* with scope for further experimentation, and with everything explained in great detail. There are also some interesting and unique features presented in this project that sets it apart from other analog designs currently residing on the Geotech website.

All the design data is open-source and freely available on the Geotech website (https://www.geotech1/forums). The source code is downloadable as an Arduino sketch, which can be accessed and modified using the Arduino integrated development environment (IDE). The electronic schematic and associated printed circuit board (PCB) design were developed using KiCad.

KiCad is an open-source software suite for Electronic Design Automation (EDA). The programs handle schematic capture and PCB layout with Gerber and NC Drill outputs, and 3D preview and export. The software suite runs on Windows, Linux and macOS and is licensed under GNU GPL v3.

The complete project was developed using tools running under the Linux Mint operating system. Both KiCad and the Arduino IDE run natively in Linux, without having to resort to the open-source Wine compatibility layer. LTspice, however, requires Wine to operate. The name *Wine* is a recursive backronym for "Wine is not an emulator". The basic operating method used in VLF metal detector technology is outlined in Chapter 1. For a deeper understanding of metal detector technology, please read *Inside the* METAL DETECTOR. (See resources and references in Appendix D.)

Chapters 1 through 12 describe the design procedures used for each stage of the project, with supporting SPICE simulations provided in Appendix C.

Chapter 13 focuses on the actual construction of the VLF metal detector. The build process is taken in stages with measurements and checks made at each stage.

Chapter 14 then explores the use of several different commercial induction balance coils.

Chapter 15 contains some final thoughts.

Appendix A contains the full component parts list - i.e. Bill of Materials (BOM).

Appendix B provides the full set of schematics.

Appendix C includes a number of SPICE simulations referred to in chapters 1 through 12.

Appendix D lists some useful resources and references.

Acknowledgements

Carl Moreland and I wrote editions 1 and 2 of *Inside the* METAL DETECTOR (often referred to as ITMD) as a collaborative effort. As a result we both learned a lot from the exercise. Previously we dedicated the book to the many contributors to the Geotech forum. I also did the same in two previous *Inner Workings Publication* books: "*The Voodoo Project*", and "*Arduino Nano Pulse Induction Metal Detector Project*".

I am particularly indebted to Carl for proofreading the material and finding numerous typos, errors, and just plain silly mistakes in the text.

After the many hours spent developing these projects, turning them into reality and creating the aforementioned books, my wife has now [*más o menos*] come to terms with my so-called "playing around with bits of wire".

Some Assumptions

To construct and set up this project, you will need a certain amount of test equipment. A multimeter and access to an oscilloscope are mandatory, and an LCR meter would be extremely useful. Plus you will need some experience with programming Arduino hardware.

Please be aware that this is not a beginner's project, as it requires a reasonable amount of electronics knowledge and some considerable patience. It is not a project that you can complete over a weekend, and it could quite possibly develop into a love/hate relationship. So be warned ...

Chapter 1   *VLF Metal Detector Technology*

"I am an electronics engineer. To save time, let's assume that I am never wrong."

--- It goes without saying

The first [obvious] question to answer is: "What does VLF mean?".

VLF is an acronym for *very low frequency*. For metal detectors this refers to transmit frequencies between 3kHz and 30kHz. Many early beat-frequency-oscillator (BFO) type detectors operated at 100kHz, as did some of the transmit-receive (TR) detectors that followed. All modern VLF metal detectors (except some dedicated to finding gold nuggets) operate in this frequency range. This is also the region we will be targetting in this project. There are many acronyms associated with metal detector technology, and to find a more in-depth explanation of these numerous terms please refer to *Inside the* METAL DETECTOR (ITMD).

Another common term is *induction balance* (IB), which refers to the coil configuration. The majority of commercial metal detectors sold today are based on the induction balance principle. Normally there is one transmit (TX) coil, and at least one receive (RX) coil, which are set up to minimize pickup between the two. When a metal target comes close to the search head, it acts like a transformer core and upsets the induction balance arrangement. The most popular coil configurations are the concentric and DD types, although you will find other variations on offer such as double-O, omega, coaxial, figure-8, and orthogonal coils, etc.

The target signal at the RX coil is extremely small and needs to be amplified before being processed by the detector's electronics. If you examine the RX signal relative to the TX signal with an oscilloscope, you will notice some interesting things. Firstly, the target needs to be close to the coil before any visual change can be identified in the scope image. Secondly, there is a phase shift introduced between the TX and RX signals which depends on the type of target. A pre-amplifier circuit, normally shortened to *preamp*, is required to boost the target signal to a useful level. The induction balance point (and the physical configuration of the coil windings) defines how the RX signal changes for ferrous and non-ferrous items, i.e., the direction of amplitude change, and whether the RX signal shifts to the left or right. It is important to understand this behaviour for each coil, although coils intended for a particular detector will be manufactured so that they all react in the same way. You will see this behaviour in more detail when we examine coils from various manufacturers in Chapter 14.

Once the incoming RX signal has been sufficiently amplified, the next step is to use a sampling technique whereby the amplitude of the received signal is measured at 0° and 90° phase positions relative to the TX signal. These two measurements are usually referred to as the X and R channels respectively. Strictly speaking, the X and R channels represent the in-phase (I) and the quadrature (Q) components of the received signal. The term *quadrature* means one quarter of a cycle - that is 90°. In other words the R channel is a measurement of the amplitude that can be attributed to the resistive aspect of the target, and the X channel is associated with the reactive part. You will see later in Chapter 6 how these signals can be used to discriminate between ferrous and non-ferrous targets, and in Chapter 11 how they can be used to provide a target ID.

GB (ground balance) is yet another acronym, which is sometimes called GEB (Ground Exclusion Balance). Sampling around the zero-crossing point of the RX signal allows the detector to ignore any changes due to the ground matrix, This technique works because the ground does not introduce a phase-shift, but only an amplitude change. Although this does assume that the ground is non-mineralized and homogeneous over a reasonable area. Shifting the GB sample slightly away from the zero-crossing point can also allow the detector to cope with moderately mineralized ground, and can either be fixed internally or, for increased flexibility, an external GB control can be provided. This is explained further in Chapters 5 and 6. As you will see, the process of sampling around the zero-crossing point is correct for Tesoro coils, but may not apply to coils from other manufacturers, but we will come to that later. There are several abbreviations used for other metal detector types, but only those mentioned above need to concern us here. For more information, please see ITMD.

For this project, a wide range of frequencies can be used. Much will depend on the actual coil that you select, but there is a huge amount of flexibility in this design. How this is actually achieved in practice will be revealed in Chapter 2.

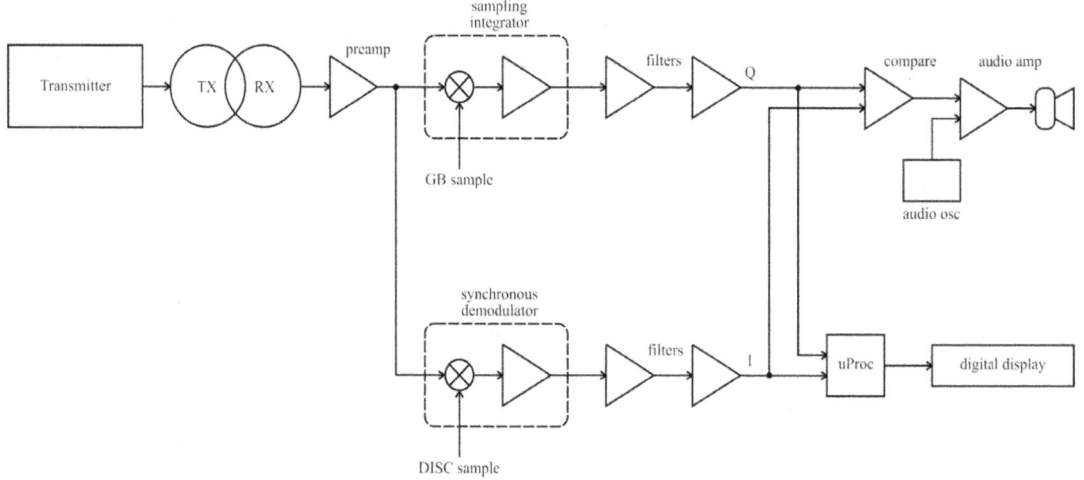

Fig. 1-1: Simplified Block Diagram of a VLF Induction Balance Metal Detector

The project presented here is a 2-channel design, where one channel is used for GB and the other for target discrimination (i.e. the rejection of undesirable metal items). Both GB and discrimination are achieved by sampling the signal from the preamp using a sampling integrator. The sampling integrator has the capability to extract a target signal from an extremely noisy environment. Subsequent filter stages manipulate the demodulated target signal using a process known as double-differentiation. The two channels are then logically combined to remove any unwanted targets, such as ferrous items. By adjusting the position of the discrimination sample, it is also possible to reject certain unwanted low conductivity items (in addition to ferrous targets) such as foil and pull tabs.

As mentioned previously, *flexibility* is a key aim for this project, and is achieved through a combination of discrete analog and digital circuits, combined with embedded software.

The signals from the GB and DISC channels are then processed by a second Arduino Nano to provide a visual display of the target ID. Any display capable of being driven by an Arduino Nano can be used, and various display types are presented in Chapter 11.

"Never trust a computer you can't throw out a window."

--- Steve Wozniak

The majority of analog metal detectors, and many that combine both analog and digital circuitry, use a transmitter in which the frequency of oscillation depends on a tuned circuit. This circuit involves a capacitor in parallel with the transmit coil in the search head. This arrangement provides a very efficient free-running oscillator, but has the disadvantage that the transmit frequency can be affected by metal targets close to the coil. Hence nearly all modern designs use a forced (or driven) oscillator where the frequency is determined elsewhere in the circuit, such as a microprocessor. In this project we are using an Arduino Nano microcontroller to generate the transmitter (TX) signal and subsequently drive the transmit coil. The transmit waveform is a sine wave, although any type of waveform (such as a square wave) could potentially be used. The main point here, as mentioned a few times already, is *flexibility*.

How then can we generate a sine wave from an Arduino Nano that does not possess a built-in digital-to-analog converter (DAC)? Obviously we could connect the Nano to an external DAC, or possibly filter a PWM output to create the sine wave. As far as the external DAC is concerned, I wanted to avoid using any special add-on boards or difficult to obtain ICs, and the PWM technique would provide less flexibility and may require component changes to enable it to work at different frequencies. In the end it was decided to use a simple inexpensive R-2R ladder DAC, which is a data converter that uses two precision resistance values to convert a digital binary number into an analog output signal proportional to the value of the digital number.

Here's how an R-2R ladder DAC works:

This method is arguably the most popular way of converting a binary digital input into an analog output. In order to understand this method, let's first determine the output voltage for a 4-bit digital input of 1000.

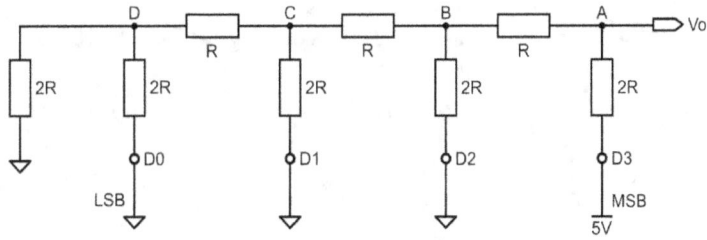

Fig. 2-1: R-2R Ladder DAC with Digital Input 1000

Referring to Fig. 2-1, the circuit shows D3 connected to 5V (logic 1) and D0 through D2 connected to 0V (logic 0). Note that the convention for binary numbers is to have the most-significant bit (MSB) on the left and the least-significant bit (LSB) on the right (which is opposite to the schematic, just to confuse things). First we need to find the equivalent resistance of the network when looking left from node A. On the left side of the ladder network at node D, you can see that we have 2R in parallel with 2R (because D0 is grounded). Hence the combination will be R, which is also in series with another R, giving 2R. That 2R is in parallel with another 2R, etc., etc. The result is that the equivalent resistance when looking left from any node will be 2R. From the equivalent circuit in Fig. 2-2, it is now immediately obvious that V_o is equal to 2.5V.

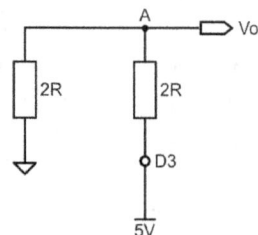

Fig. 2-2: Equivalent Circuit for R-2R Ladder DAC with Digital Input 1000

If we follow the same procedure for a digital input of 0100 (Fig. 2-3), then D2 is connected to 5V and inputs D0, D1, and D3 are connected to 0V. As before, the equivalent resistance when looking left from node B will be 2R, and the equivalent circuit is shown in Fig. 2-4. Then looking left into the network from node A we see that the voltage B will be 2.5V, and V_o will be 1.25V.

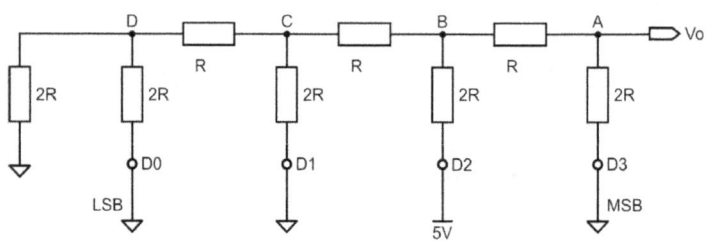

Fig. 2-3: R-2R Ladder DAC with Digital Input 0100

By the same analysis we can determine that for 0010, $V_o = 0.625V$, and for 0001, $V_o = 0.3125V$. In general:

$$V_o = \frac{5V}{2^{N-n}}$$

[Eq.1]

where N = the number of binary inputs

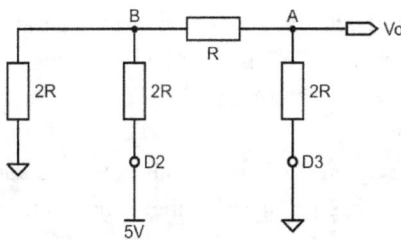

Fig. 2-4: Equivalent Circuit for R-2R Ladder DAC with Digital Input 0100

For any combination of binary inputs, V_o can be calculated by using the principle of superposition by simply adding the voltages together. For example, 0110 will result in $V_o = 1.25 + 0.625 = 1.875V$.

Table. 2-1 shows the value of V_o for all possible binary combinations. Note that V_o cannot reach the maximum value of 5V, but is limited to 4.6875V. V_o should also be connected to a high impedance buffer in order to provide an accurate conversion. Typical values for R and 2R are 10k and 20k respectively.

Binary Inputs				V_o
D3	D2	D1	D0	(V)
0	0	0	0	0.0000
0	0	0	1	0.3125
0	0	1	0	0.6250
0	0	1	1	0.9375
0	1	0	0	1.2500
0	1	0	1	1.5625
0	1	1	0	1.8750
0	1	1	1	2.1875
1	0	0	0	2.5000
1	0	0	1	2.8125
1	0	1	0	3.1250
1	0	1	1	3.4375
1	1	0	0	3.7500
1	1	0	1	4.0625
1	1	1	0	4.3750
1	1	1	1	4.6875

Table. 2-1: All Possible Binary Combinations for a 4-bit R-2R Ladder DAC

This is all very well, but how does this help us to generate a sine wave using the Arduino Nano?

With the Arduino programming language it is a simple matter to generate the correct binary inputs for the R-2R ladder DAC. The full program will be discussed in Chapter 4, but let's examine how this is achieved.

The Arduino Nano needs to output a sequence of binary values that equate to a sine wave, and we can get the Nano to calculate these values for itself as part of the setup routine. We only need to generate the required values for one complete cycle, as the waveform will repeat ad infinitum. In this design we will be using an 8-bit R-2R ladder DAC, which will provide 256 amplitude steps, as opposed to the 16 steps for the 4-bit version. This will also allow V_o to reach 4.9805V.

First we set the degree variable to zero.

```
deg = 0;
```

Then step the variable i for 240 steps, because we want to use increments of 1.5° (and 240 * 1.5 = 360°).

```
for (int i = 0; i <= 239; i++) {
```

Convert degrees to radians.

```
rad = deg / 360 * TWO_PI;
```

Calculate the value of sine for the current value of degree, and then scale and offset by 2.5V.
This will produce a sine wave that has a peak-to-peak value of 5V centred around 2.5V.

```
tx = sin(rad) * 2.5 + 2.5;
```

Since this is an 8-bit R-2R ladder, there will be 256 binary values, giving a DAC resolution of 19.6mV. Next store the calculated values in a one-dimensional array.

```
dacInput[i] = tx / 0.0196;
```

Finally, increment degrees by 1.5.

```
deg += 1.5;
```

And repeat for all 240 steps.

```
}
```

The stored values are then output on one of the digital ports of the Nano and applied to the binary inputs of the R-2R ladder DAC (see Chapter 4).

In order to provide a powerful enough output to drive the TX coil, the DAC output needs to be amplified further. In Fig. 2-5 you can see that the DAC output is fed into a push-pull output stage. The +12V supply comes directly from the battery, and the -12V supply is generated by U3, which is a voltage converter (LT1054) described in the power supply section of the next chapter. The sine wave output from the TX amplifier is shown in Fig. 2-6.

The TX circuit was also simulated using LTspice. The full details of this simulation can be found in Appendix C. For the simulation, a transmit frequency of 7.5kHz was arbitrarily chosen, but of course other transmit frequencies are possible, as we will see later.

Before we proceed further, it is worth pointing out some pin restrictions associated with the Arduino Nano, and an often overlooked discrepancy between the Uno and the Nano.

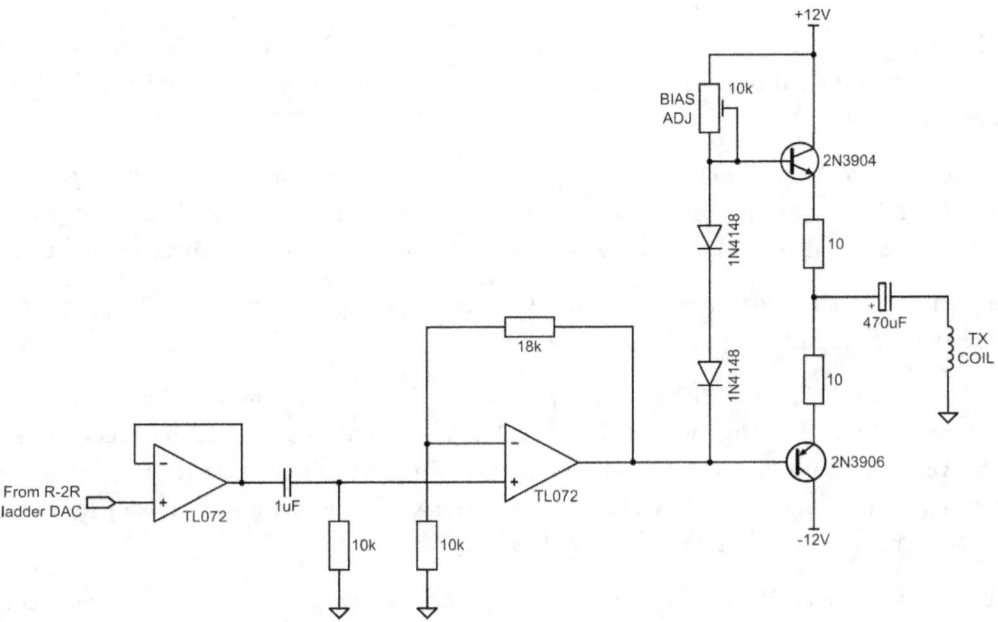

Fig. 2-5: TX Amplifier Circuit

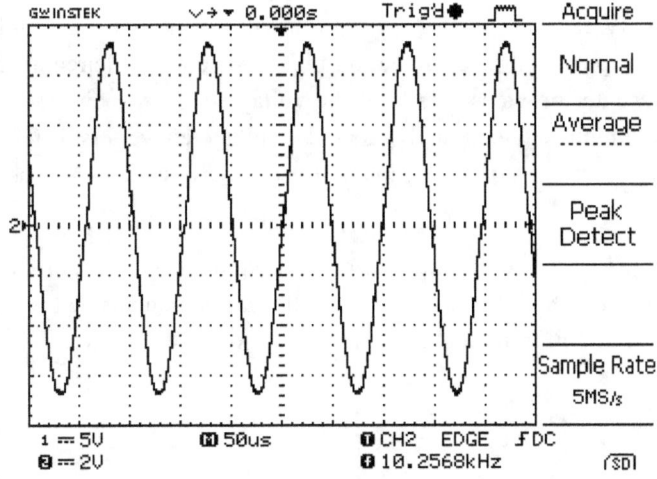

Fig. 2-6: TX Amplifier Output

Arduino Nano Port Limitations

D0 and D1 are used by the internal hardware UART, and are therefore not readily available for use. Although these digital ports can be used (with some restrictions) it prevents the use of Serial.print() for debugging purposes.

D13 is attached to an onboard LED, and it is recommended that it only be used as an output. If it is unavoidable that D13 has to be assigned as an input, it requires an external pull-down resistor. The internal pull-up resistor should not be used. A6 and A7 on the Arduino Nano can only be used as inputs.

The above points are rarely made clear, and there are numerous people experiencing problems and asking the same questions in the Arduino forums.

There is a discrepancy between the D0 and D1 pinouts for the Arduino Uno and the Nano. This can become an issue if you use an Uno in conjunction with a prototyping board to test out some ideas, and then switch over to a Nano. It is also further confused by some Nano pinout diagrams on the Internet which incorrectly show pin 1 as D0/RX and pin 2 as D1/TX. This is the correct pinout for the Uno, whereas the Nano has pin 1 as D1/TX and pin 2 as D0/RX, as shown in Fig. 2-7.

Note: In the Arduino Nano PI Metal Detector Project, I was caught out by both an erroneous online pinout diagram and an incorrectly connected Arduino Nano symbol in the DipTrace library. This resulted in a mistake in the schematic on page 9 (Fig. 2-1) where D0 and D1 were swapped over. Luckily this did not affect correct operation of the circuit since these pins were unused.

An Arduino Nano is used to generate the TX oscillator signal, the audio tone, and the trigger signals for the sample pulses required by the synchronous demodulators. It is also used to provide the +5V supply used by the analog circuitry. If you examine Fig. 2-7 you will see that a +12V supply is connected to VIN on pin 30, and the +5V supply is then available on pin 27. The analog circuitry requires +5V and -5V supplies, and the next chapter provides details of the voltage converter circuitry. In addition, the Nano generates a synchronization pulse to externally clock the voltage converters. If the voltage converters are allowed to free-run (that is, to use their own internal oscillators) switching transients could potentially affect the RX signal.

Virtually any induction balanced search head can be used with this design, with the proviso that it contains one TX coil and one RX coil, without any additional components that would prevent the TX coil being force driven or the RX coil from being tuned with an external capacitor.

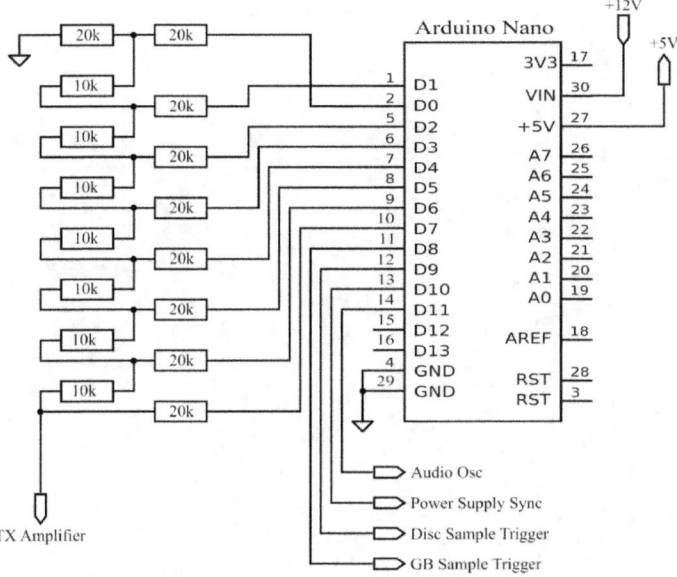

Fig. 2-7: Transmitter Oscillator

Chapter 3

Voltage Converters

"Is this a diode which I see before me? The anode toward my hand? Come, let me clutch thee. I have thee not, and yet I see thee still."

--- Macbeth misquoted

Although there is already +12V available from the battery pack, and a stabilised +5V supply provided by the Arduino Nano, there is an additional requirement for -12V and -5V supplies.

The TL072 opamp in the TX amplifier is powered from +/-12V, as is the amplifier output stage. This is provided from the upper LT1054 in Fig. 3-1, which is a switched-capacitor voltage converter IC, often referred to as a "negative voltage generator" or "negative charge pump". The negative output is -Vcc, but with an inevitable small voltage drop. In practice, with an input voltage of +12V the output is more likely to be closer to -11.4V.

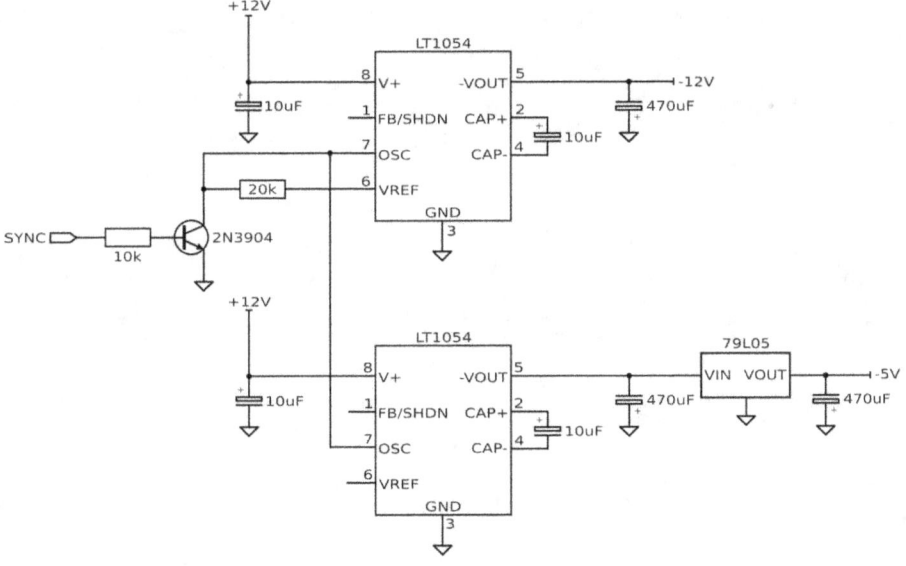

Fig. 3-1: Voltage Converters

The lower LT1054 in Fig. 3-1 performs the same function of generating a -12V output, which is then regulated by a 79L05 linear voltage regulator to provide a stabilised -5V output. There are two voltage converters used in this design in order to split the power budget between the two, and to reduce any interaction between the TX amplifier stage and the other more sensitive analog circuitry.

The Arduino Nano also provides a synchronization pulse (SYNC) to externally clock both voltage converters. Note (from Fig. 2-7 and Fig. 3-1) that port D10 of the Nano is connected to the converters via an NPN transistor whose collector is pulled up by a 20k resistor to an internal voltage reference of +2.5V nominal. The external clock allows any noise from the converters to become synchronized to the sampling pulses, and thereby eliminate any switching transients that could potentially affect the receive signal. The SYNC pulse is generated by the Arduino software, which is described in detail in the next chapter.

"If builders built buildings the way programmers wrote programs, then the first woodpecker that came along would destroy civilization."

--- Weinberg's second Law

The Arduino project started life in 2005, and was developed as an educational tool for students at the Interaction Design Institute Ivrea in Ivrea, Italy. It was intended to be an inexpensive and easy way for novices and professionals to create devices that could interact with their environment through measurement and control. Arduino uses open-source hardware, and the Arduino Integrated Development Environment (IDE) is freely available under _copyleft_ licenses. The microcontroller boards are pre-programmed with a bootloader that makes it very easy to upload programs to the on-board flash memory.

Most of the microcontroller I/O pins are accessible via headers on the Arduino boards. The Nano board used in this design has two male headers that connect to the PCB and provide the necessary timing signals and inputs required by the detector.

The Arduino IDE is a cross-platform application for Windows, macOS, and Linux.

This particular project was completely developed using the Linux Mint (version 20.3 Una) operating system with the Cinnamon desktop environment (version 5.2.7).

Also (for reference) the hardware used was a Dell Vostro 3525 laptop with a 6-core AMD Ryzen 5 5625U CPU, Radeon Graphics (1920x1080), 15.5" screen, 16GB RAM, and 512GB SSD. Although Linux Mint 20.3 Uma will run comfortably with a much lower specification 64-bit machine having 1024x768 screen resolution, 4GB RAM, and 100GB disk space.

It is assumed that you have previously been using the Arduino IDE to explore the capabilities of the device, and already know how to upload software to the microcontroller's flash memory.

Programs written with the Arduino IDE are known as a _sketch_, and are saved as text files with the extension ".ino". Arduino programs consist of two functions, _setup()_ and _loop()_. The _setup()_ function is called once when a sketch starts after power-up or reset. It is used to initialize variables, input and output pin modes, and other libraries needed in the _sketch_. It is analogous to the function _main()_. After the _setup()_ function completes, the _loop()_ function executes continuously and is analogous to the function _while(1)_.

The *sketch* for the Arduino Nano VLF Detector starts with a title comment.

```
// Arduino Nano VLF metal Detector
// Nano #1 with TX sine wave output
```

Next we need to assign variables to any microcontroller pins. In this case, we only need to assign a variable for the audio tone output, which is allocated to port D11. The other outputs are handled in a manner that does not require a variable assignment. This will become clear later.

```
// Pin assignments
byte audioPin = 11;        // Audio tone
```

Several variables are required by the program to allow calculation of the data points for the sine wave output, plus one to record the current state of the SYNC pulse for the voltage converters.

```
// Program variables
float deg, rad;            // Degrees and radians
float tx;                  // TX value for each data point
byte dacInput[240];        // R-2R ladder DAC input values
byte syncState = LOW;      // State of SYNC pulse (HIGH or LOW)
```

Then we need a routine to toggle the SYNC pulse.

The function *digitalWrite()* is commonly used to control digital output on the Arduino pins. The use of *digitalWrite()* adds a level of abstraction that separates the software from the physical pin, and does not require the programmer to know anything about the underlying hardware. On the downside, *digitalWrite()* (on the Nano) takes 3.6μs to determine which pin is mapped to the selected port and whether the pin can be used for hardware PWM (if so, this needs to be turned off), then finally create the bit mask for the pin. All of this takes time. The advantage is that the code is cross-compiler compatible and easy to understand, but is relatively slow when compared to accessing the ports directly.

For our purposes, controlling the SYNC pulse using *digitalWrite()* causes a potential problem. The SYNC pulse is toggled at the start of each TX cycle, which produces a discontinuity in the TX sine wave output. To minimize any discontinuity it is preferable to toggle the SYNC pulse as quickly as possible by directly accessing the pin. The selected output is set HIGH using a logical-OR bit mask, and set low using a logical-AND bit mask.

For comparison, a direct-access write takes 120ns as opposed to 3.6μs for a *digitalWrite()* operation.

However, before we can use the direct-access method, we need to understand how the ATmega328 processor pins map to the Arduino Nano ports. This is shown in Table. 4-1.

Since the SYNC pulse is output on port D10 (pin 13) on the Nano (see Fig. 2-7), it is clear from Fig. 4-1 that the ATmega328 refers to this pin as PB2 (i.e. PORT B2).

```
// Power supply sync pulse (D10)
void sync() {
  if (syncState == LOW) {
    PORTB |= B00000100;  // If SYNC state is LOW, then toggle B2 HIGH
    syncState = HIGH;    // Save current SYNC state
  } else {
    PORTB &= B11111011;  // If SYNC state is HIGH, then toggle B2 LOW
    syncState = LOW;     // Save current SYNC state
  }
}
```

The same technique of direct-access write to the port pins is also used to generate the ground balance (GB) and discrimination (DISC) sample trigger pulses. The GB trigger is output on PORT B0 (D8), and the DISC trigger is output on PORT B1 (D9). Again, refer to Fig. 4-1.

```
// GB sample trigger (D8)
void gbTrigger() {      // Output GB trigger pulse
  PORTB &= B11111110;   // Set B0 LOW
  PORTB |= B00000001;   // Set B0 HIGH
}

// DISC sample trigger (D9)
void discTrigger() {    // Output DISC trigger pulse
  PORTB &= B11111101;   // Set B1 LOW
  PORTB |= B00000010;   // Set B1 HIGH
}
```

As mentioned previously, the *setup()* function is always executed first following power-up or reset.

First we need to calculate the DAC input values for one complete cycle of the TX waveform, and the most obvious question that comes to mind is: How many steps are required for the sine wave?

By a process of empirical measurement (commonly referred to as *trial and error*) it was determined that 240 steps of 1.5° would create a sine wave with a frequency of 10.25kHz. Other frequencies can easily be generated to suit the coil under test, as you will see in Chapter 14.

Digital ports D0 to D11 are set to output mode. Note (from Fig. 2-7) that D0 to D7 are all connected to PORTD on the ATmega328 processor.

For the TX oscillator and output stage to start up correctly the -12V supply must be available. Since the voltage converters rely on the existence of a SYNC pulse, this is output for a period of one second in the *setup()* routine to allow the voltage converters to achieve a steady-state condition. After that the interrupts are turned off to prevent jitter in the TX waveform. Disabling interrupts will stop the *delay()* function from working, but this is not a problem because it is only used in the *setup()* routine to kick the voltage converters into life.

Lastly, the audio oscillator output is configured for PWM (490Hz).

```
void setup() {
  // Calculate DAC input values
  deg = 0;                                  // Start with zero degrees
  for (int i = 0; i <= 239; i++) {          // Sine wave will consist of 240 steps
    rad = deg / 360 * TWO_PI;               // Convert degrees to radians
    tx = sin(rad) * 2.5 + 2.5;              // Scale and offset
    // 5V divide by 255 = 0.0196
    // R/2R ladder DAC resolution is 8-bits (19.6mV)
    dacInput[i] = tx / 0.0196;              // Save value for current step
    deg += 1.5;                             // Increment by 1.5 degrees
  }
  for (int i = 0; i <= 11; i++) {           // Set D0 to D11 as outputs
    pinMode(i, OUTPUT);
  }
  for (int i = 0; i <= 10000; i++) {        // Drive SYNC pulse for 1 second
    sync();
    delayMicroseconds(100);
  }
  noInterrupts();                           // Disable interrupts
  analogWrite(audioPin, 127);               // Set audioPin with 50% duty cycle PWM
}
```

The *loop()* function first calls the *sync()* function before applying the positive half of the sine wave to the DAC inputs. One thing that was not mentioned in the discussion on direct-access write is that it is also possible to switch more than one bit at the same time. This is not possible with a *digitalWrite()* operation. Since the R-2R ladder DAC inputs are all connected to PORTD on the ATmega328, then we only have to point PORTD to the correct location in the array that contains the sine wave values.

The application of the sine wave is performed in two parts because the GB and DISC trigger pulses need to occur before the negative half of the sine wave. This is a requirement for the test coil, and is discussed further in Chapters 6 and 14.

```
void loop() {
  sync();                                   // Sync voltage converters
  for (int i = 0; i <= 119; i++) {          // Apply positive half of sine wave to DAC inputs
    PORTD = dacInput[i];
  }
  gbTrigger();                              // GB sample trigger
  discTrigger();                            // DISC sample trigger
  for (int i = 120; i <= 239; i++) {        // Apply negative half of sine wave to DAC inputs
    PORTD = dacInput[i];
  }
}
```

That's it for the software in Arduino Nano #1. The Arduino Nano #2 software will be dissected in Chapter 12.

For now we'll put our programmer's hat on one side and proceed to look at the rest of the hardware in the next chapter.

Arduino Nano		ATmega328	
Pin Number	Pin Label	Pin Number	Pin Label
1	TX1	31	PD1
2	RX0	30	PD0
3	RST	29	PC6
4	GND	-	-
5	D2	32	PD2
6	D3	1	PD3
7	D4	2	PD4
8	D5	9	PD5
9	D6	10	PD6
10	D7	11	PD7
11	D8	12	PB0
12	D9	13	PB1
13	D10	14	PB2
14	D11	15	PB3
15	D12	16	PB4
16	D13	17	PB5
17	3V3	-	-
18	REF	21	AREF
19	A0	23	PC0
20	A1	24	PC1
21	A2	25	PC2
22	A3	26	PC3
23	A4	27	PC4
24	A5	28	PC5
25	A6	19	ADC6
26	A7	22	ADC7
27	5V	-	-
28	RST	29	PC6
29	GND	-	-
30	VIN	-	-

Table. 4-1: Arduino Nano to Atmega328 Pin Map

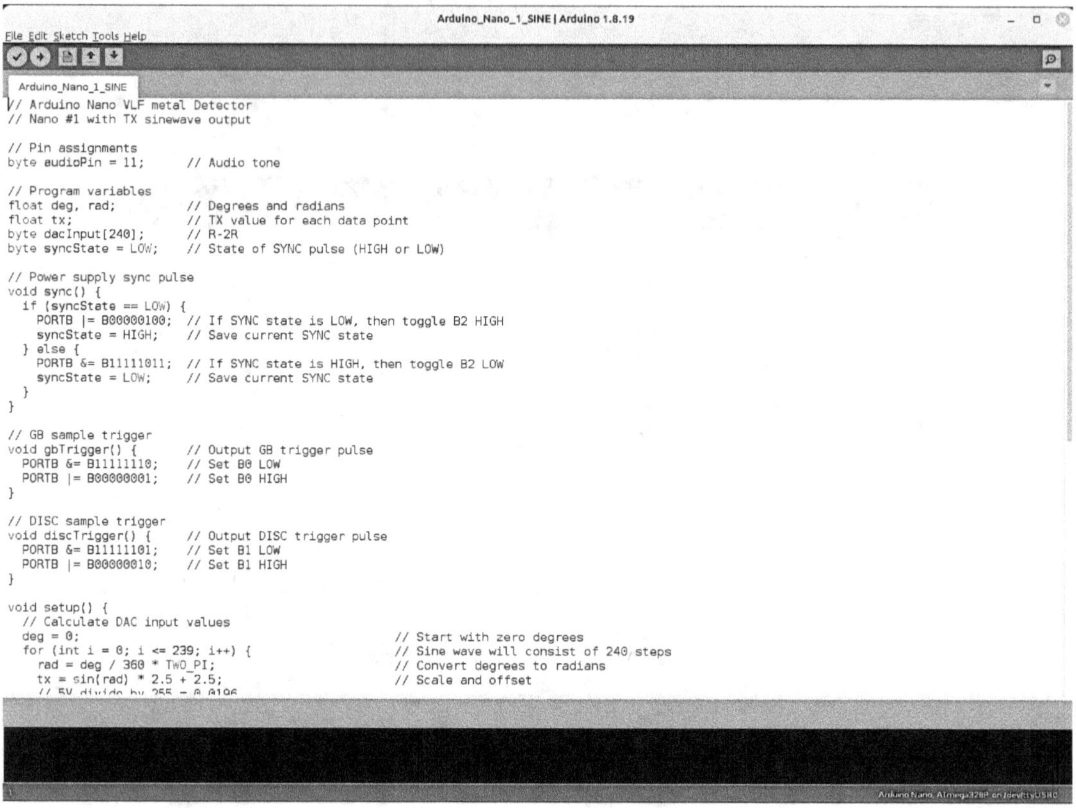

File Edit Sketch Tools Help

Arduino_Nano_1_SINE

```
// Arduino Nano VLF metal Detector
// Nano #1 with TX sinewave output

// Pin assignments
byte audioPin = 11;       // Audio tone

// Program variables
float deg, rad;           // Degrees and radians
float tx;                 // TX value for each data point
byte dacInput[240];       // R-2R
byte syncState = LOW;     // State of SYNC pulse (HIGH or LOW)

// Power supply sync pulse
void sync() {
  if (syncState == LOW) {
    PORTB |= B00000100;   // If SYNC state is LOW, then toggle B2 HIGH
    syncState = HIGH;     // Save current SYNC state
  } else {
    PORTB &= B11111011;   // If SYNC state is HIGH, then toggle B2 LOW
    syncState = LOW;      // Save current SYNC state
  }
}

// GB sample trigger
void gbTrigger() {        // Output GB trigger pulse
  PORTB &= B11111110;     // Set B0 LOW
  PORTB |= B00000001;     // Set B0 HIGH
}

// DISC sample trigger
void discTrigger() {      // Output DISC trigger pulse
  PORTB &= B11111101;     // Set B1 LOW
  PORTB |= B00000010;     // Set B1 HIGH
}

void setup() {
  // Calculate DAC input values
  deg = 0;                                    // Start with zero degrees
  for (int i = 0; i <= 239; i++) {            // Sine wave will consist of 240 steps
    rad = deg / 360 * TWO_PI;                 // Convert degrees to radians
    tx = sin(rad) * 2.5 + 2.5;                // Scale and offset
    // 5V divide by 255 = 0.0196
```

Arduino Nano, Atmega328P on /dev/ttyUSB0

Fig. 4-1: Arduino Integrated Development Environment (IDE)

"All truths are easy to understand once they are discovered, the point is to discover them."

--- Galileo Galilei (1564-1642)

The basic preamp circuit is shown in Fig. 5-1. Note that the receive (RX) coil requires a parallel connected tuning capacitor that is dependent on the inductance of the coil being used and the TX frequency. There will be a residual signal from the transmit circuit of a few mV present at the RX coil, which can be seen by connecting an oscilloscope, and by bringing various metallic objects close to the coil this signal will change in amplitude and also in phase relative to the TX signal. However, these changes are quite small and need to be amplified and further processed to extract the faint target signal.

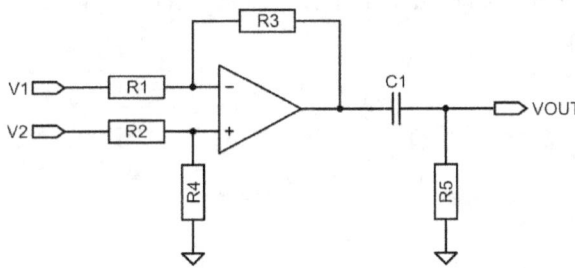

Fig. 5-1: Preamp Circuit

With reference to Fig. 5-1, note that the preamp is configured as a differential amplifier so that it acts like both an inverting and non-inverting amplifier at the same time, with the result that it subtracts one input from the other.

One question that may come to mind is how can the subtraction process be very accurate when the inverting configuration gain for an opamp is:

$$G_{inv} = \frac{R_3}{R_1}$$

[Eq.1]

whereas the non-inverting configuration is:

$$G_{ninv} = 1 + \frac{R_3}{R_1}$$

[Eq.2]

If we connect the non-inverting input in Fig. 5-1 to ground, we will end up with the configuration shown in Fig. 5-2. In this case the output voltage is given by:

$$VOUT_{inv} = -V_1(\frac{R_3}{R_1})$$

[Eq.3]

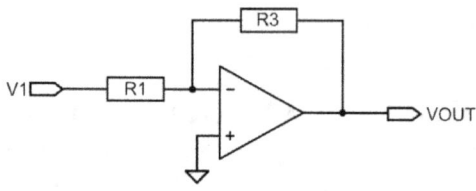

Fig. 5-2: Inverting Opamp Configuration

On the other hand, if we short the V_1 side of R_1 to ground and inject the input signal directly into the non-inverting input, we will have the standard non-inverting amplifier configuration shown in Fig. 5-3, which has an output voltage given by:

$$VOUT_{ninv} = V_3(1 + \frac{R_3}{R_1})$$

[Eq.4]

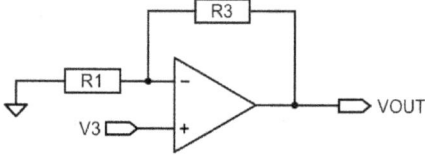

Fig. 5-3: Non-inverting Opamp Configuration

The voltage at the non-inverting input in Fig. 5-1 can be found by using the Potential Divider Rule:

$$V_3 = V_2(\frac{R_4}{R_2 + R_4})$$

[Eq.5]

Substituting Eq.5 into Eq.4, we have:

$$VOUT_{ninv} = V_2(\frac{R_4}{R_2 + R_4})(1 + \frac{R_3}{R_1})$$

[Eq.6]

To find V(out) for the differential amplifier in Fig. 5-1, we can use the principle of superposition and simply add the VOUT$_{inv}$ and VOUT$_{ninv}$ voltages together, as follows:

$$VOUT = -V_1(\frac{R_3}{R_1}) + V_2(\frac{R_4}{R_2 + R_4})(\frac{R_1 + R_3}{R_1})$$ [Eq.6]

If we make $R_1=R_2$ and $R_3=R_4$, the equation becomes:

$$VOUT = -V_1(\frac{R_3}{R_1}) + V_2(\frac{R_3}{R_1 + R_3})(\frac{R_1 + R_3}{R_1})$$ [Eq.6]

which reduces to:

$$VOUT = -V_1(\frac{R_3}{R_1}) + V_2(\frac{R_3}{R_1})$$ [Eq.7]

The differential amplifier transfer function is therefore:

$$VOUT = (V_2 - V_1)(\frac{R_3}{R_1})$$ [Eq.8]

As you can readily see, the output voltage from the differential amplifier circuit is simply the difference between the two inputs multiplied by the ratio of the feedback to input resistor, and the different gains for the two configurations is not a problem.

You may also have looked at Fig. 5-1 and wondered why we should use a differential amplifier anyway, particularly since (in most cases) the RX coil has one side connected to ground. In fact, simply looking at the schematic could imply that the R_2 and R_4 potential divider will be shorted to ground. In reality, these two ground connections may be separated by some distance with an inevitable potential difference between them. Tesoro coils, for example, have their RX ground connection to the cable shield within the coil shell. The preamp is designed to only amplify differential signals (as can be seen from Eq.8) and cancels any common-mode signals. With the Tesoro coils, any noise that is common to both of the RX wires will be cancelled by the preamp. Regardless of where the coil shield is connected to ground, the differential preamp is able to handle all scenarios.

At the output of the preamp (Fig. 5-1) there is a passive high-pass filter. The combination of a 220nF capacitor and a 1k resistor provides a cutoff frequency of:

$$f_c = \frac{1}{2\pi RC} = \frac{1}{2\pi * 220nF * 1k} = 723Hz$$ [Eq.9]

which will effectively reduce any interference from the mains (either 50Hz or 60Hz, depending on where you live). The series capacitor also blocks any dc offset introduced by the preamp.

Chapter 6 _Synchronous Demodulators_

"Not everything that counts can be counted, and not everything that can be counted counts."

--- Sign hanging in Einstein's office at Princeton

The process of synchronous demodulation extracts the target signal. In a 2-channel VLF detector there are two samples required: the GB sample and the DISC sample.

For Tesoro coils the GB sample occurs at, or close to, one of the zero-crossing points on the RX waveform. Which zero-crossing point depends on the coil being used, as the presence of a metal target needs to increase for the design presented here. The reason for sampling around the zero-crossing point is to effectively ignore any signal due to the ground which manifests itself as a change in amplitude without any phase-shift. Homogeneous mineralized ground can be handled to a degree by sampling slightly away from the zero-crossing point, although heavily mineralized soil generally requires more complex solutions.

On the other hand, the DISC sample occurs at a point 90° away from the zero-crossing at the peak of the RX waveform. At this sample point on the RX waveform, non-ferrous targets produce an increase in the demodulator voltage, and ferrous targets produce a decrease. Again, depending on the particular coil, the DISC sample could be on either the positive or negative half cycle. This design is flexible enough to handle any of these scenarios, once you've tested the coil in question to determine how it reacts.

In order to demodulate the target signal we need to provide some means of sampling the received signal and storing the result until the next cycle. Since the GB and DISC sample pulses are synchronized to the TX signal, this is why it's known as _synchronous demodulation_. This technique is also referred to as _lock-in detection_. In essence synchronous demodulators use the knowledge about a signal's time dependence to extract it from a noisy background.

With reference to Fig. 6-1, the GB and DISC samples pulse their respective CMOS analog switch. The 20k resistor, CMOS switch, and 100nF output capacitor combine to act as a simple sample-and-hold circuit which is capable of storing the result until the next sample pulse. With the GB sample pulse positioned at the zero-crossing point of the RX signal, the output of the analog switch should be zero volts. If you adjust the GB control to either side of this position, this will cause the amplitude to rise or fall. For the GB channel, the output should increase for both ferrous and non-ferrous targets. The DISC sample pulse must be located at the peak of the receive signal, which may either positive or negative depending on the coil. In this case a non-ferrous target will cause an increase in voltage and a ferrous target will cause the voltage to decrease. In practice (if setting up the detector to have a fixed ground balance) the GB sample will need to be offset slightly from the zero-crossing point in order to ignore a small ferrite tuning slug. Such a target is generally agreed to represent low ground mineralization. If an external GB control is included, then the user can ground balance more accurately for the actual ground conditions.

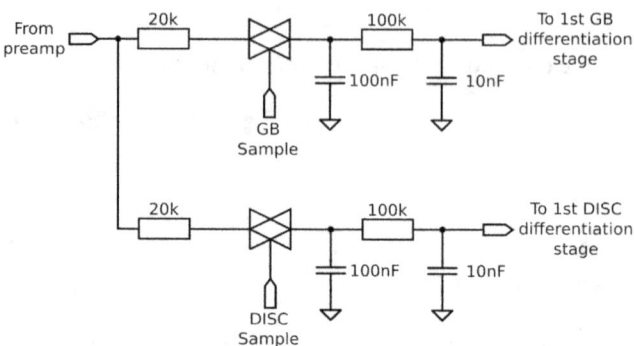

Fig. 6-1: Synchronous Demodulators

At the output of both sample-and-hold circuits is an additional low-pass RC filter with a cut-off frequency of 159Hz in order to remove any remnants of the TX frequency from the sampled outputs.

"A common mistake that people make when trying to design something completely foolproof is to underestimate the ingenuity of complete fools."

--- Douglas Adams

The ground balance (GB) and discrimination (DISC) channels need to be demodulated to extract the target signal from the preamp output. In order to accomplish this task it is necessary to generate both GB and DISC sample pulses to clock the synchronous demodulators (described in Chapter 6). There are two possible ways to do this. The first might be to generate the sample pulses directly from the Arduino Nano under software control, and the second method would be to do this with hardware.

The software approach not only requires the generation of the delays and sample pulses, but also the ability to read the external GB and DISC controls and adjust the delays accordingly. During testing, the software was controlled by a state machine with the external controls being read periodically after several TX cycles. It soon became clear that there were a number of problems associated with this approach. Firstly, it was necessary to ensure that the GB and DISC samples were never allowed to overlap, which required the sampling to take place on alternate cycles. Secondly, there was significant interruption of the TX signal whenever the GB and DISC controls were read. Thirdly, this interruption required a sequence of at least 5 non-sampled cycles so that the RX signal could return to a steady-state condition.

The final conclusion was that this was not a viable approach using the Arduino platform, and it was decided to proceed along the hardware route.

Figures 7-1 and 7-2 show the sample pulse generation circuitry for the GB and DISC channels respectively.

The Arduino Nano #1 generates both trigger pulses for each channel (as shown in Fig. 2-7) which are used to initialize the first monostable on the negative-going edge of the trigger. This first monostable generates a pulse at the Q output with a width defined by the external resistor and capacitor combination. CD14538BE dual-monostable ICs are used in this design, so that the pulse width is simply RC.

For example, with C=10nF, and R=5k, the pulse width will be 50uS.

This first monostable sets the delay from when the trigger pulse occurs and the start of the actual sample pulse. The start of the trigger pulse is controlled by software, and the minimum delay is set by an onboard preset. The user is also provided with an optional external GB control for adjustment in the field. If the external control is not fitted, the detector will operate with a _factory-set_ GB setting.

When the first monostable is triggered, the Q output goes high for the defined period, and triggers the second monostable on the falling edge. The output of the second monostable is taken from the ~Q output and generates a sample pulse that goes low during the sample pulse period. This inversion is necessary because there is a transistor connected to the output which inverts the pulse and converts it from a 5V pulse to +/-5V to be compatible with the sample gates.

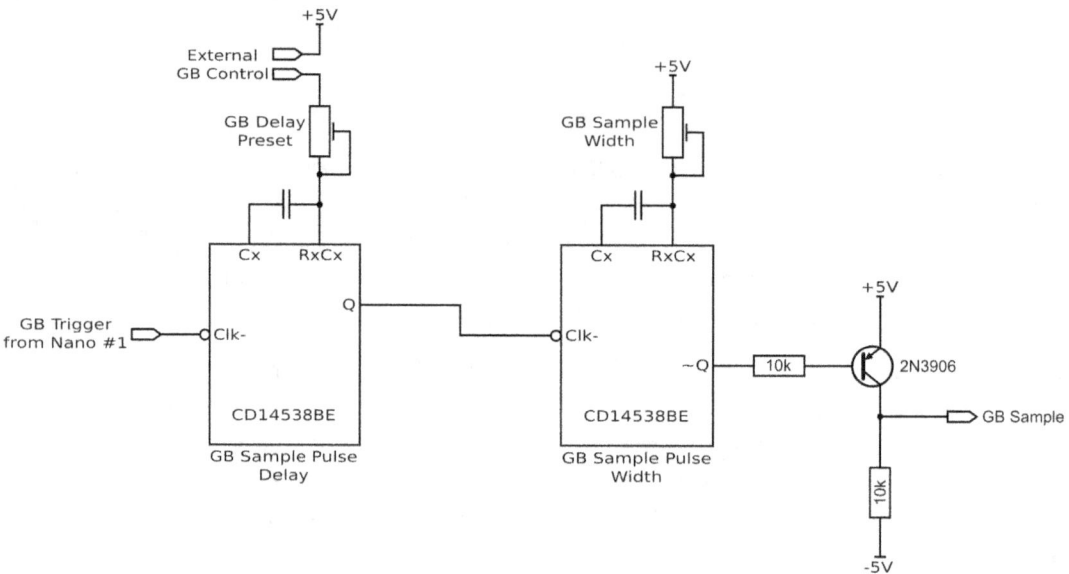

Fig. 7-1: GB Sample Pulse Generator

Using the Arduino software to define the trigger point for the GB and DISC pulses, and then hardware to create the delays and the pulses themselves, provides much greater flexibility with regards to using coils from different manufacturers. Virtually any permutation can be catered for with a little experimentation, as you will see in Chapter 14.

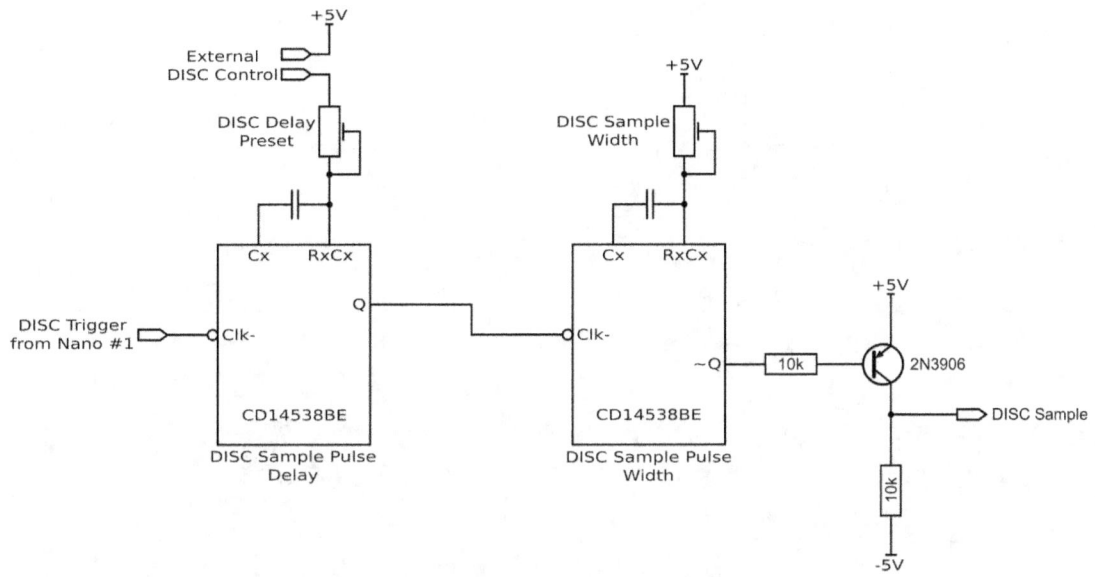

Fig. 7-2: DISC Sample Pulse Generator

"What I will need from you now is a commitment. You will listen closely, and you will not judge me until I am finished. If you cannot commit to that, then please leave the room. But if you choose to stay, remember you chose to be there. What happens from this moment forward is not my responsibility."

--- Alan Turing (The Imitation Game) - 2014

The VLF metal detector presented in this project, like virtually all modern designs, is intended to be a motion detector. In other words it should respond to the rate-of-change of the signal, rather than the signal itself. The theory is that any ground signal changes will be relatively slow, whereas any target signals will produce a fast change as they pass through the electromagnetic field of the coil. This technique also helps to ignore any slow changes caused by component or temperature drift.

To react to the target signal only when it's changing, we need to differentiate the output of the synchronous demodulators. A differentiator produces an output that is the derivative of the input, and a practical differentiator circuit is shown in Fig. 8-1.

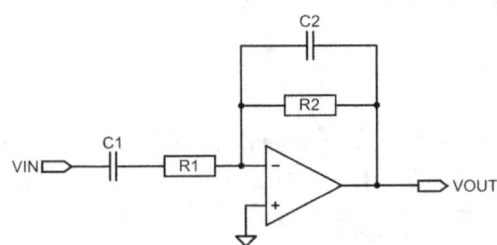

Fig. 8-1: Practical Differentiator

R_1 and C_1 define the break frequency (f_b) of the circuit, and R_2 and C_2 define the cutoff frequency (f_c). The cutoff frequency is important because the differentiator amplifies the input signal in direct proportion to the frequency. Without C_2 there would be a lot of high frequency noise at the output, and the combination of R_2 and C_2 introduces a cutoff frequency that is below that of the upper cutoff frequency of the opamp.

Common values for f_b and f_c are around 7Hz and 15Hz respectively, with the gain of the 1st differentiator being 100x (40dB).

So how do we calculate the required component values?

The break frequency (f_b) can be calculated using Eq.1:

$$f_b = \frac{1}{2\pi R_1 C_1}$$

[Eq.1]

If we arbitrarily set R_1 to 47k, then:

$$C_1 = \frac{1}{2\pi f_b R_1} = \frac{1}{2\pi \times 7 \times 47k} = 484nF$$

Choosing the nearest preferred value of 470nF for C_1 gives a break frequency of:

$$f_b = \frac{1}{2\pi \times 47k \times 470n} = 7.2Hz$$

And for a gain of 100x:

$$R_2 = R_1 \times 100 = 47k \times 100 = 4M7$$

A resistance for R_2 of 4M7 is quite a high value for the feedback resistor, but ignoring that for the moment, let's calculate a suitable value for C_2 to achieve a cutoff frequency (f_c) close to 15Hz:

$$C_2 = \frac{1}{2\pi f_c R_2} = \frac{1}{2\pi \times 15 \times 4M7} = 2.25nF$$

Choosing the nearest preferred value for C_2 of 2n2 gives a cutoff frequency of:

$$f_c = \frac{1}{2\pi \times 4M7 \times 2n2} = 15.4Hz$$

[Eq.2]

Although we could juggle around with the component values in an attempt to reduce the value of R_2, let's explore an interesting alternative that involves the use of a complex network in the feedback path.

In general, these complex feedback networks can be difficult to analyse because the usual solution involves solving the node or loop equations. Also, you cannot use the superposition theorem because there is only one input voltage. The simple solution is to use Thevenin's Theorem. The use of a T-network allows more flexibility by using a low dc resistance path in the feedback loop.

For example, examine the circuit in Fig. 8-2:

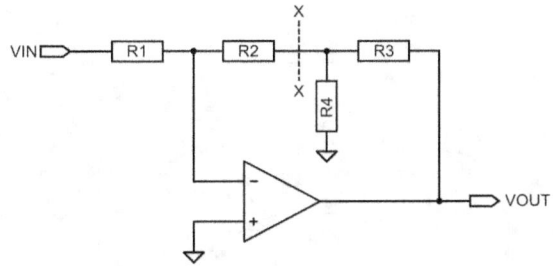

Fig. 8-2: Complex Feedback Network

If we break the circuit at X-X, we can then calculate the Thevenin voltage and impedance by using the divider rule:

$$V_{TH} = V_{out} \frac{R_4}{R_3 + R_4} \qquad \text{[Eq.3]}$$

where V_{TH} is the no-load voltage, and R_{TH} (parallel combination of R_3 and R_4) is the impedance seen looking into the network

Now let's redraw the circuit with Thevenin equivalents (Fig. 8-3).

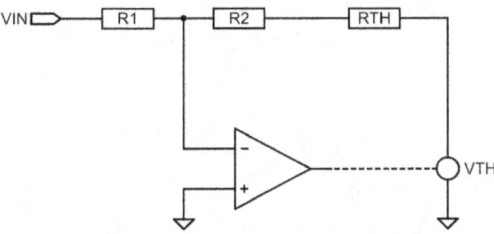

Fig. 8-3: Thevenin Equivalent Circuit

The gain of the circuit is:

$$G = \frac{V_{TH}}{V_{in}} = \frac{R_2 + R_{TH}}{R_1} \qquad \text{[Eq.4]}$$

Then, replacing V_{TH} with V_{out} (from Eq.3] gives:

$$\frac{V_{out}}{V_{in}} = \frac{R_2 + R_{TH}}{R_1} \times \frac{R_3 + R_4}{R_4}$$

and replacing R_{TH} with $R_3 \parallel R_4$:

$$\frac{V_{out}}{V_{in}} = \frac{R_2 + (R_3||R_4)}{R_1} \times \frac{R_3 + R_4}{R_4} \qquad \text{[Eq.5]}$$

Since the gain of the circuit is given by:

$$\frac{V_{out}}{V_{in}} = \frac{R_f}{R_1}$$

Then it is clear from Eq.5 that the equivalent feedback resistor must be:

$$R_f = [R_2 + (R_3||R_4)] \times \frac{R_3 + R_4}{R_4} \qquad \text{[Eq.6]}$$

The parallel combination of R_3 and R_4 is:

$$R_3||R_4 = \frac{R_3 R_4}{R_3 + R_4}$$

Substituting for ($R_3 || R_4$) in Eq.6 allows us to reduce the equation further:

$$R_f = \frac{R_2(R_3 + R_4) + R_3 R_4}{R_4}$$

$$= \frac{R_2 R_3 + R_2 R_4 + R_3 R_4}{R_4}$$

$$= \frac{R_2 R_3}{R_4} + \frac{R_2 R_4 + R_3 R_4}{R_4}$$

Resulting in a simple equation for the equivalent feedback resistor R_f that corresponds to the complex network:

$$R_f = \frac{R_2 R_3}{R_4} + R_2 + R_3 \qquad \text{[Eq.7]}$$

At the moment we have the circuit shown in Fig. 8-1. However, a feedback resistor of 4M7 is not really practical, so we need to replace this with a T-network.

For our purposes in this design we need to use the Thevenin Theorem in reverse, as we already know the value for the single feedback resistor and wish to convert to this into a T-network.

First rearrange Eq.7 to obtain an expression for R_4:

$$R_4 = \frac{R_2 R_3}{R_f - R_2 - R_3} \qquad \text{[Eq.8]}$$

We know that $R_f = 4M7$, and if we set R_2 and R_3 to 470k:

$$R_4 = \frac{470k \times 470k}{4M7 - 470k - 470k} = 58.75k$$

Using the nearest preferred value of 56k will result in an equivalent feedback resistor of:

$$R_f = R_2 + R_3 + \frac{R_2 R_3}{R_4} = 470k + 470k + \frac{470k \times 470k}{56k} = 4M88$$

and a gain of:

$$G = \frac{R_f}{R_1} = \frac{4.88M}{47k} = 103.8 \ (40.3dB)$$

Combined with the capacitor value shown of 2n2, the break frequency of the differentiator is fb = 7.2Hz, and the cutoff frequency fc = 14.8Hz.

The complete differentiator circuit using a complex T-network in the feedback loop is shown in Fig. 8-4.

A comparison between the standard differentiator and the version with the T-network is performed in a SPICE simulation in Appendix C.

To prevent the outputs of the synchronous demodulators being loaded by the first differentiator stage, we need to make a small change to the final circuit. This involves changing the topology of the opamp from inverting to non-inverting. This change affects the gain of the circuit, which is increased by 1 (see Chapter 5, Eq.2).

The final circuit is shown in Fig. 8-5.

At the beginning of this chapter it was stated that the detector should respond to the rate-of-change of the signal. This is not exactly accurate as it really needs to respond to the (rate-of-change)^2 of the signal. This process, known as double-differentiation, provides a sharp response to metal targets while drastically reducing any response to the ground matrix. For an in-depth explanation of the differentiation process, please see ITMD.

Providing the detector with the ability to double differentiate is simply a matter of adding a second differentiator circuit. This second circuit usually has a lower gain of around 45x (33dB) so a complex feedback network will not be necessary in this instance. Fig. 8-6 shows the second filter stage. Using the equations for the practical differentiator presented earlier you should be able to confirm that the break frequency (f_b) is 7.2Hz, the cutoff frequency (f_c) is 15.9Hz, and the gain (G) is 45x (33dB).

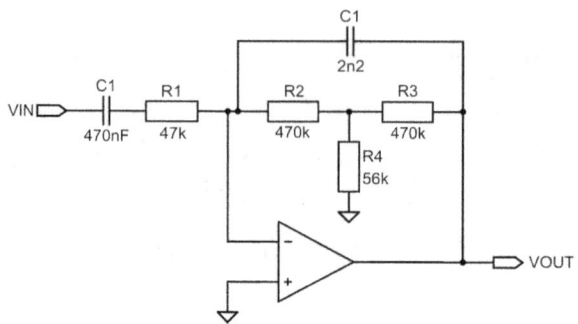

Fig. 8-4: Differentiator Circuit with T-network

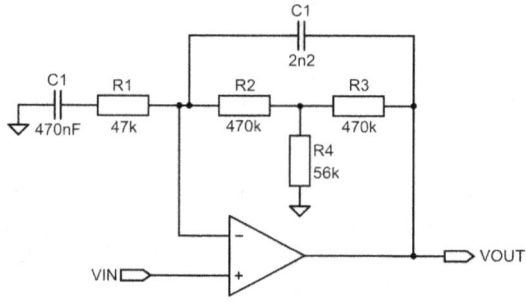

Fig. 8-5: Differentiator Circuit with T-network (Non-inverting Topology)

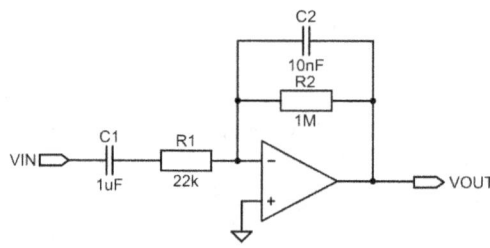

Fig. 8-6: Second Filter Stage

The differentiating filter stages are identical in both the GB and DISC channels. If you trace the target signal through the different parts of the circuitry, with the knowledge that the GB demod output increases for all targets, then it may appear at first sight that the audio tone will decrease. However, you have to account for the fact that the double-differentiation (X") process inserts an extra *apparent* inversion in the chain. This is because the response, after double-differentiation takes place, has 3 peaks. Without signal inversion there would be 2 beeps for every non-ferrous target, and no signal as the target passed over the centre of the coil. This is more clearly shown in Fig. 8-7, where X" is the double-differentiated target response. As you can see, the double-differentiated response (X") has its "main peak" inverted.

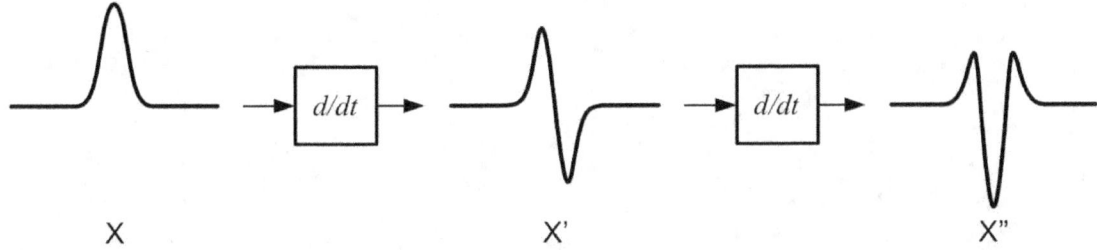

$$X \qquad\qquad X' \qquad\qquad X"$$

Fig. 8-7: Raw (X), 1ˢᵗ derivative (X'), 2ⁿᵈ derivative (X")

As mentioned previously, the output of the GB channel should increase for all targets, whereas the DISC channel will increase for non-ferrous targets, and decrease for ferrous. Adjustments can be made to the default GB sampling delay to eliminate moderately mineralized ground, and the same for the DISC channel to reject foil, bottle caps, pull-tabs, etc. In chapters 11 and 12 we will be discussing how to provide a visual indication of the possible target.

Chapter 9 _____ *Comparators*

"There is a theory that states that if anyone discovers what the Universe is for and why it is here, it will instantly disappear and be replaced by something even more bizarre and inexplicable. There is another theory that this has already happened."

--- Douglas Adams

In order to boost the weak signals from deep targets, and hence increase sensitivity, comparators can be placed between the GB and DISC channel outputs and the audio amplifier. See Fig.9-1. Without the addition of the comparators the amplitude of the audio output will depend on the strength of the target signal, which means that you have to listen very carefully for those weak targets. The comparator circuit effectively boosts all target signals to the same level. The main advantage is an increased detection distance, but with the downside that there is no longer any audio indication of depth. However, the target ID display (described in Chapter 11) provides a solution to this problem.

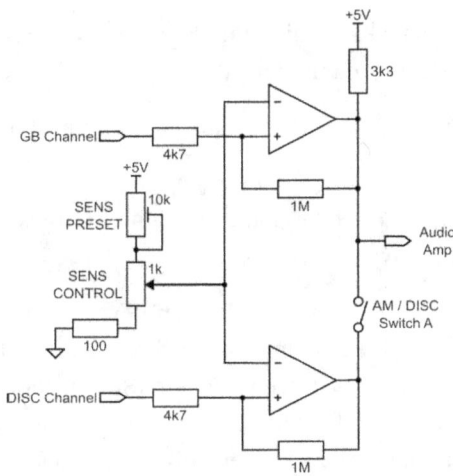

Fig. 9-1: Comparators

The simplest way to explain the operation of the comparator circuit is to refer to a truth table (Table. 9-1).

A '1' in the table indicates that the particular output signal (from either the GB or DISC channel) is a logic high (+5V) and a '0' indicates that the signal is a logic low (-5V). From the table it is immediately clear that the AUDIO output is only enabled when both the GB and DISC channels are high. This indicates that a non-ferrous target has been detected.

A closer examination of the table will reveal exactly how this works:

STATE	DISC	GB	AUDIO
0	0	0	0
1	0	1	0
2	1	0	0
3	1	1	1

Table. 9-1: Comparator Truth Table

State 0: When no target is being detected, both GB and DISC channel outputs will be low, and therefore there will be no audio.

State 1: Remember that the GB channel goes high for all targets, whereas the DISC channel goes high for non-ferrous and low for ferrous. Hence, in this state, the target must be ferrous and there is no audio.

State 2: The DISC channel is high, but the GB channel is low indicating that there is no target present. Again, no audio.

State 3: Both GB and DISC channels are high, indicating a non-ferrous target detection. This is the only combination that can produce an audio response.

At first sight, state 2 may seem to be a situation that can never occur in practice. Due to the double-differentiation process, the GB channel first swings low before it goes high, and then swings low a second time. It does this for every target, whatever the type. The DISC channel does the exact same thing for a non-ferrous target, but the reverse occurs for ferrous. In this case state 2 represents the situation when a ferrous target is detected and the logic state represents the side lobes on either side of the main signal, which (of course) means that they are ignored.

As the truth table shows, the comparator circuit acts like a logic AND gate.

The comparator IC used in this design is an LM393 dual package. These comparators have an open-collector output which requires a pull-up resistor, and allows the outputs to be connected together to effectively provide the logic AND function.

In Fig. 9-1 you can see that there are 1M resistors providing positive feedback. The combination of the 4k7 and 1M resistors are there to implement hysteresis and thereby prevent chatter in the output caused by instability due to noise. This is a common problem with comparators when the input and reference voltages are almost equal. Instead of having one simple threshold, hysteresis can be used to set an upper and lower threshold to eliminate multiple transitions caused by noise. The amount of hysteresis in Fig. 9-1 can be calculated as follows:

$$Hysteresis = \frac{R_1}{R_2}(V_{OH} - V_{OL}) = \frac{4k7}{1M}(5 - (-5)) = 47mV$$

To ensure that the output from the comparators do not invert the signal from the GB and DISC channels, they need to be configured in the non-inverting mode. In this configuration the hysteresis is applied to the incoming GB and DISC channels rather than the reference threshold.

The Sensitivity (SENS) control is externally adjustable by the user to set the audio threshold. There is also an internal SENS Preset which sets a limit on how high the sensitivity can be adjusted.

The switch in Fig. 9-1 enables the user to select either All Metal (AM) or Discrimination (DISC) modes. When the switch is open, this disconnects the DISC comparator output allowing the audio to respond to all metal targets regardless of type.

"How is education supposed to make me feel smarter? Besides, every time I learn something new it pushes some old stuff out of my brain. Remember when I took that wine-making course, and I forgot how to drive?"

--- Homer Simpson

The audio tone is generated by the Arduino Nano #1 on port D11, which is configured for pulse width modulation (PWM). This generates a digital rectangular wave with a constant frequency and a duty cycle of 50%.

Configuring a PWM output on the Arduino is a simple matter of calling: *analogWrite(pin, value)*

where (in this case) *pin* = 11, and *value* = 127.

Note that there is no need to call *pinMode()* to set the output before calling *analogWrite()*. The parameter *value* can vary from 0 (always off) to 255 (always on) which defines the duty cycle, and a value of 127 will set the duty cycle to 50%.

On the Nano (by default) pins 3, 9, 10 and 11 can output a frequency of 490Hz, and pins 5 and 6 can output 980Hz when configured for PWM. The ATmega328 processor has 3 PWM timers that are used to control 6 PWM outputs. Although it is possible to manipulate the chip's timer registers directly to obtain more control, the default frequency on pin 11 of 490Hz is suitable for our purposes in this design.

Referring to the audio amplifier output stage shown in Fig. 10-1, the transistor that is driven by the audio oscillator is often called a *chopper* since it "chops" the bias signal. The audio output is biased on by the outputs from the GB and DISC channels. When a metal target is detected, the bias signal increases and turns on the first transistor allowing the audio signal to pass through to the final audio stage that drives the speaker or headphones.

The lower transistor is not actually part of the amplifier stage, but is mentioned here since it is in the same circuit block. It generates a "Target Detection" signal which informs the Arduino Nano #2 that (as the names suggests) a metal target has been detected. This information is used to drive the selected display (see Chapter 11).

As shown in the previous chapter, the comparator outputs can only go high when a metal target has been detected. In AM mode, this means any metal target. In DISC mode, this means only non-ferrous targets. If the comparator outputs are low, then the cathode of the diode will also be low and the audio output is disabled. As soon as the comparator outputs go high, the audio tone from the oscillator is allowed through to the speaker or headphones to provide the user with an audible target response.

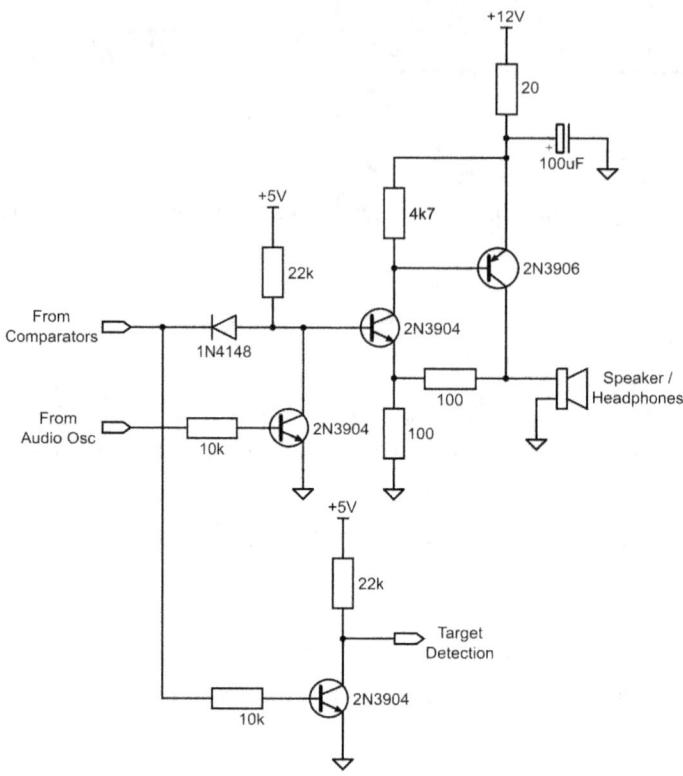

Fig. 10-1: Audio Amplifier

"'Forty-two', said Deep Thought, with infinite majesty and calm."

--- The Hitchhiker's Guide to the Galaxy

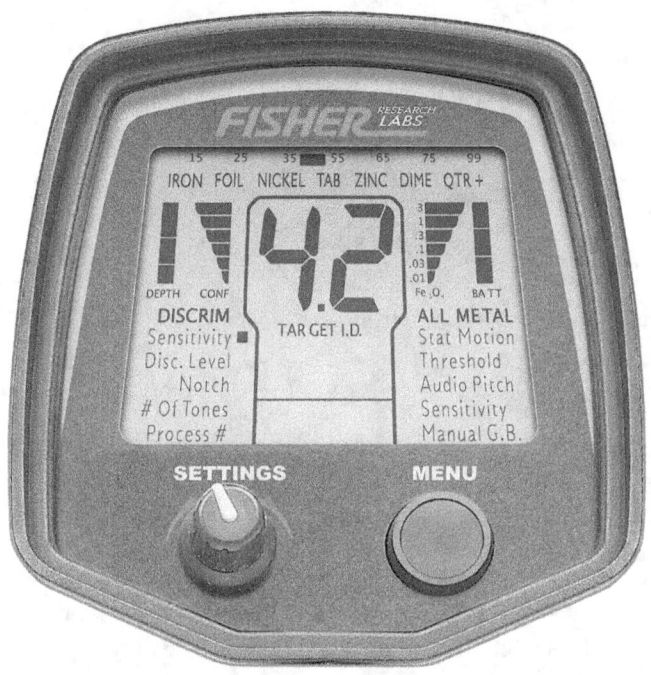

Fig. 11-1: The Meaning of Life, the Universe, and Everything

Unlike Deep Thought's very precise answer to "What is the meaning of life, the universe, and everything?", a metal detector's ability to calculate the phase angle associated with a buried target is not an exact science. The presence of junk, ground mineralisation, and the depth of the target can make identification somewhat *fuzzy*, and the fact that certain objects exhibit different responses when the coil is swept across them in different directions doesn't help either. Nonetheless, a target ID can be a very useful feature on a metal detector.

The process by which a target ID is derived is known as phase demodulation. As outlined in Chapter 6, a sampling technique is employed whereby the received signal is measured at the ninety and zero degree phase positions. The measurements are either referred to as the in-phase (I) and quadrature (Q) samples, or the reactive (X) and resistive (R) parts of the received signal. The R-channel is used to achieve ground balance, as it is the one that has quadrature sampling. The X-channel is clocked in-phase and therefore sees the full reactive ground signal. Since a ferrite ideally has a 0° phase shift you can see from the equations below that X = A and R = 0, meaning that R is balanced. The differentiating nature of the channel filters also reduces sensitivity to any slow amplitude changes which helps to ignore ground response.

Phase demodulation is applied to the outputs of the GB and DISC channels by using some basic trigonometry. Mathematically the demodulated signals are:

$$X = A \cos\theta$$

and

$$R = A \sin\theta$$

Since

$$\tan\theta = \left(\frac{\sin\theta}{\cos\theta}\right)$$

then

$$\tan\theta = \left(\frac{R}{X}\right)$$

Consequently, the phase angle θ can be found from

$$\theta = \tan^{-1}\left(\frac{R}{X}\right)$$

This all sounds more complicated than it is in practice, as you will see in the next chapter.

Some detectors, like the Fisher F75, display the target ID as well as a moving cursor which suggests what the target may actually be. In Fig. 11-1 the ID is *42*, and the bar shows that it could be a U.S. Nickel or a pull-tab. Of course, it might be something completely different. The only way to be 100% sure is to dig it, as it could also be a Roman coin. At the end of the day, the target ID is just a useful guide.

It is also worth noting that the target ID system of numbers is not consistent across different metal detector manufacturers. A target that displays as *42* on a Fisher F75 (for example) may be different on another detector. The general consensus for mainstream manufacturers though appears to be that target IDs are scaled to a range between 0 and 99.

The phase discrimination method measures the effective electrical conductivity of the probable metal target. This depends on the composition of the object, as well as its size, shape, and orientation in the soil. Coins are the most easily identified due to their consistency. Smaller or thinner objects all have a lower conductivity, as do those made of iron, bronze, lead, pewter, or zinc. Larger objects, and those made of silver, copper, and aluminium will give a higher conductivity reading. Gold is the exception. Although you might think that gold would have a high conductivity value, this is not the case for gold nuggets, as they are rarely found in large pieces. Ferrous (iron) targets can also confuse the target ID system if they are ring-shaped, and some flat pieces of iron or steel, such as can lids, may also be misidentified.

In general, a metal detector will be able to make an intelligent guess at the majority of targets in an air test to a distance of approximately 10 inches (25cm). This reduces to about 8 inches (20cm) in the ground.

Since we don't have easy access to a sophisticated custom display such as the one shown in Fig. 11-1, what are the alternatives?

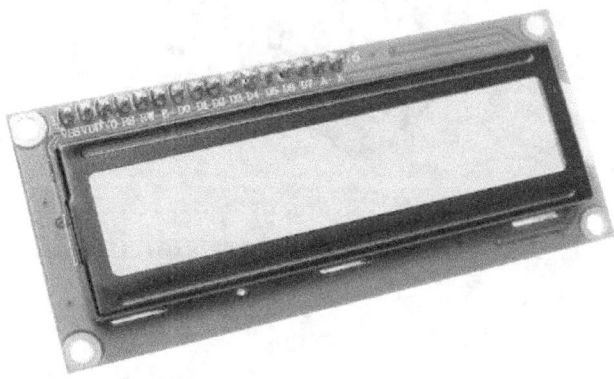

Fig. 11-2: 2-line 16-character LCD

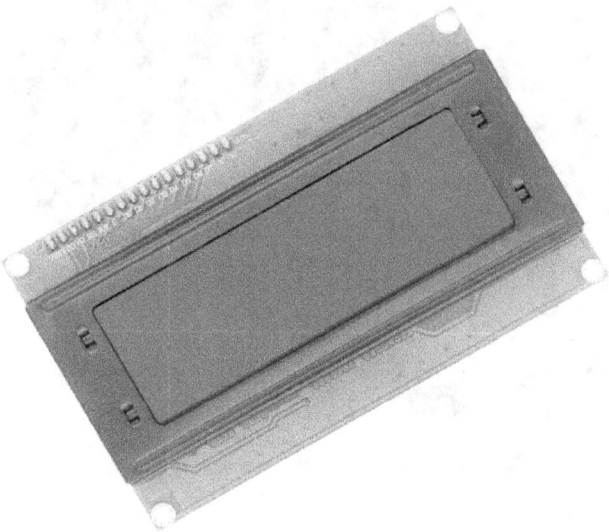

Fig. 11-3: 4-line 16-character LCD

The 2-line 16-character LCD shown in Fig. 11-2 is one of the easiest displays to use with the Arduino Nano, but is hardly one of the most exciting. However, it is quite capable of displaying the necessary information relating to a metal target.

If we don't want to be restricted to 2-lines of 16-characters, then the 4-line display shown in Fig. 11-3 is a possible alternative, and is just as simple to control using the same library functions.

Fig. 11-4: SH1106 OLED 1.3" (128x64) Display

Fig. 11-5: TFT 2.4" (320x240) Display

The OLED in Fig.11-4 provides a very clear and bright display, but the disadvantage here is the small size at only 1.3".

Like the OLED, the TFT display in Fig. 11-5 can display graphics as well as text. In the following chapter we will be experimenting with all of these displays to check out their pros and cons.

It is important to note that the coordinate system for graphical displays is for 0,0 to be at the top left-hand corner of the screen. Positive X increases to the right, and positive Y increases downwards. This is upside down when compared to the standard Cartesian coordinate system. This method originates from the days of cathode ray tube (CRT) systems where the raster started in the top left-hand corner and scanned across and down the screen.

Interesting Fact

The Arduino IDE's built-in font is based on the original IBM PC character set, known as Code Page 437 (or CP437 for short). When CP437 was first transcribed into the GFX library, one symbol was accidentally left out, with the result that the GFX library is off by one. Unfortunately it is now too late to correct the error. As a compromise solution, the original IBM CP437 symbol values can be enabled using:

display.cp437(true);

By default the erroneous version is enabled by the Arduino IDE.

So now I hear you asking: "OK – but which symbol is missing?", and the answer is: 0xB0 (decimal 176). See Fig. 11-6.

The outputs from the GB and DISC channels are fed into a resistive potential divider referenced to +5V. This results in any ADC reading above 512 as being positive, and below as negative. From these signals it is possible to calculate the target phase as described earlier, and also the signal strength. How this is achieved will be revealed in the next chapter.

Referring to Fig. 11-7, you can see that connector J11 provides connections for the LCD and TFT displays, and J13 is reserved for the OLED. In addition, to provide further flexibility, any unused ports are broken out on J10 in case they may be useful during experimentation. It should be obvious that J11 can be connected to either an LCD or a TFT, but not both at the same time. Note that (for some inexplicable reason) I left out D3 when breaking out the unused ports.

The LCD displays are driven using a 4-wire parallel interface, whereas the TFT uses the serial peripheral interface (SPI). The OLED is assigned to J13 because it uses a multi-controller / multi-target bus known as an I2C interface. You may sometimes see this referred to as I^2C or IIC.

When using an LCD you will inevitably find versions that have different pin numbering. It is becoming normal practice for the pins to be numbered from 1 to 16 (which is just common sense, you would think), but another common variant has its pins numbered 15, 16, and then 1 to 14. Why this is so is anybody's guess. A third variant (which is now much less common) has two rows of 8 pins, with an additional requirement for -5V to drive the backlight.

A reflective LCD uses the ambient light in the vicinity. The light passes through the LCD layer to a mirror, which reflects it back to the user. Viewing can therefore be difficult in dark rooms or outdoors at night. Transmissive LCDs require a backlight, otherwise the screen is blank. By far the most popular LCDs are transflective, as they use both transmissive and reflective methods.

CP437

	_0	_1	_2	_3	_4	_5	_6	_7	_8	_9	_A	_B	_C	_D	_E	_F
0_	NUL	SOH	STX	ETX	EOT	ENQ	ACK	BEL	BS	HT	LF	VT	FF	CR	SO	SI
1_	DLE	DC1	DC2	DC3	DC4	NAK	SYN	ETB	CAN	EM	SUB	ESC	FS	GS	RS	US
2_	SP	!	"	#	$	%	&	'	(	)	*	+	,	-	.	/
3_	0	1	2	3	4	5	6	7	8	9	:	;	<	=	>	?
4_	@	A	B	C	D	E	F	G	H	I	J	K	L	M	N	O
5_	P	Q	R	S	T	U	V	W	X	Y	Z	[	\	]	^	_
6_	`	a	b	c	d	e	f	g	h	i	j	k	l	m	n	o
7_	p	q	r	s	t	u	v	w	x	y	z	{	|	}	~	DEL
8_	Ç	ü	é	â	ä	à	å	ç	ê	ë	è	ï	î	ì	Ä	Å
9_	É	æ	Æ	ô	ö	ò	û	ù	ÿ	Ö	Ü	¢	£	¥	Pts	ƒ
A_	á	í	ó	ú	ñ	Ñ	ª	º	¿	⌐	¬	½	¼	¡	«	»
B_	░	▒	▓	│	┤	╡	╢	╖	╕	╣	║	╗	╝	╜	╛	┐
C_	└	┴	┬	├	─	┼	╞	╟	╚	╔	╩	╦	╠	═	╬	╧
D_	╨	╤	╥	╙	╘	╒	╓	╫	╪	┘	┌	█	▄	▌	▐	▀
E_	α	β	Γ	π	Σ	σ	µ	τ	Φ	Θ	Ω	δ	∞	φ	ε	∩
F_	≡	±	≥	≤	⌠	⌡	÷	≈	°	∙	·	√	ⁿ	²	■	NBSP

Fig. 11-6: IBM CP437 Character Set

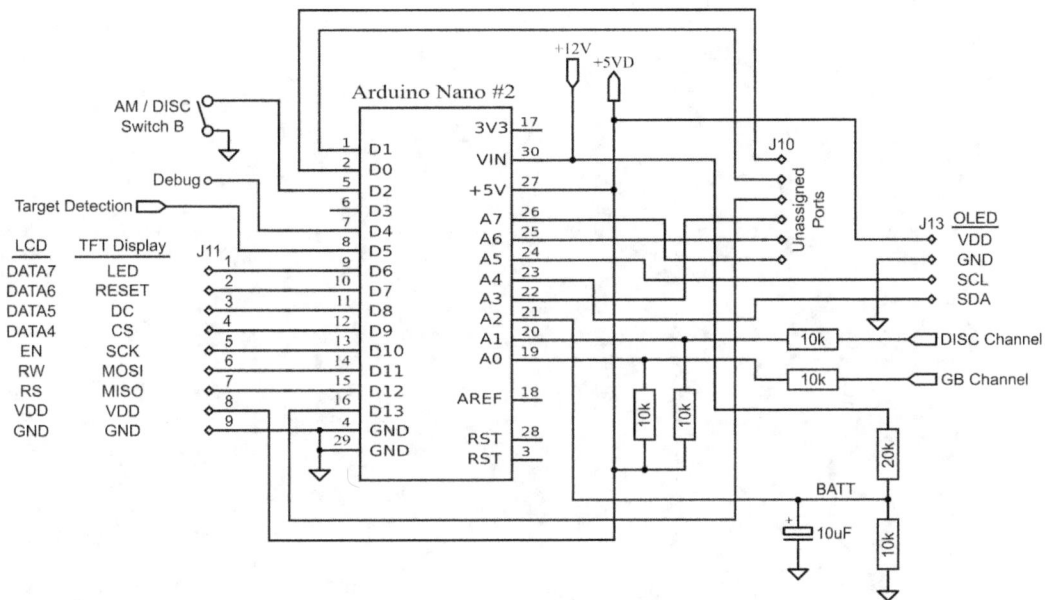

Fig. 11-7: Arduino Nano #2

As well as the target ID and signal strength, the Arduino Nano #2 uses one of its ADC ports to measure the battery voltage and display the state of charge. The maximum voltage that can be read by the Nano's ADC ports is 5V, and the value returned by the *analogRead()* function ranges from 0 to 1023. Hence the smallest voltage that can be measured is 4.88mV. Since the battery voltage is divided by 3 by the resistive potential divider, then a value of 1 represents a voltage of 14.64mV, and 1023 represents 14.98V. We will be discussing battery pack options in more detail in Chapter 13.

In Fig. 9-1, AM / DISC is connected to part-A of a double-pole single-throw (DPST) switch, and in Fig. 11-7 AM-DISC is connected to part-B. This allows the second Arduino Nano to know when the detector is operating in either All-Metal or DISC modes, and to adjust the display parameters accordingly.

"Oh … I forgot the semicolon."

--- Every software engineer

The details of how to connect the various displays is given in Chapter 13 – Step 19. This chapter deals with the software for each of the displays mentioned in Chapter 11.

Before we get to the actual code for each display, there are a number of operations that are common to both the 2-line and 4-line LCDs. To avoid repetition these will be described first, pointing out any differences between the two LCDs.

As usual, the Arduino _sketch_ starts with a comment that provides a summary of the operation of the software. In order to communicate with an alphanumerical liquid crystal display (LCD), we have to include the _LiquidCrystal.h_ library. This library is based on the Hitachi HD44780 (or compatible) chipset, which is found on most text-based LCDs. The library works using either 4 or 8 bit modes in addition to the RS, EN, and RW control lines.

The LCD connections are included as comments for reference, and the necessary LCD pin assignments are then made. Both the 2-line and 4-line LCDs are operated in 4-bit mode.

Specific variables are then assigned to various microcontroller pins. Note that a debug pin is also assigned to assist with software debugging, and to time certain events in conjunction with an oscilloscope.

Next we need to declare some useful program constants.

The _sampleDelay_ constant is used to introduce a delay between when the target trigger is initiated and the point where the Nano samples the signal at the outputs of the GB and DISC channels. The required value for this delay can be determined by setting the _debug_ pin high after the target trigger is detected, and resetting it after the specified number of samples have been taken. With the correct delay, the _debug_ pulse should be aligned with the positive peak of the double-differentiated channel outputs. This is also important for obtaining accurate VDI numbers.

The display counter limit of 2000 provides an approximate time of 3 seconds for the target VDI and signal strength to remain on the display before being automatically erased. An initial solution using interrupts was tested to provide this functionality, but was eventually replaced with a much simpler solution of counting the number of program loops during which no target was detected.

The purpose of the declared variables should be fairly obvious once you read the comments. The trigonometrical arctangent function performs its calculations in radians, which is why the code includes variables for both radians and degrees. For display purposes we need the phase angle in degrees.

A special function is included for measuring the battery voltage. The Nano's ADC has a resolution of 4.88mV, and a potential divider reduces the battery voltage to bring it within the range of the ADC. Hence we have to multiply the ADC reading by 4.88E-3 * 3 to get the actual battery voltage.

For the 2-line LCD, the setup routine loads some custom characters into the CG RAM of the LCD. These characters are displayed dependent on the state of the battery pack. When the battery voltage drops below 10.5V, the battery icon changes to an X to indicate that the pack needs recharging. The custom characters in the 2-line LCD version are used to represent the charge state of the battery due to limited space, but with the 4-line LCD it is possible to display the actual battery voltage, and the custom characters are not used. The measured and scaled voltage is first multiplied by 10 and then converted to an integer. For example, a value of 125 would represent 12.5V. Each digit is subsequently extracted using the Arduino div and mod commands to display the voltage to one decimal place at the end of line 3, as shown in Fig. 13-33 (next chapter).

The *setup()* routine is always executed first following power up or reset. Initially the pin modes are defined. In this case we only need to define the *debugPin* as an output. The AM/DISC switch input is analog, and does not need to be defined since this is the default mode. However, we do need to specify that the input requires the internal 20k pullup resistor to be enabled. When the AM/DISC switch is on, the input is pulled low.

The next part of the *setup()* routine initializes the LCD by displaying a splash screen announcing the detector is an "Arduino Nano VLF Metal Detector". This message is displayed for 3 seconds, and any target information on the screen is cleared.

For the 2-line LCD a number of custom characters are then defined and loaded into the CG RAM of the LCD. These custom characters show the battery in various states of charge. There is one other custom character for a non-filled block that is used to indicate the target ID on a VDI bar in the second row of the LCD. For the 4-line LCD the only custom character created is the non-filled cursor block. This is also used to indicate the signal strength of the target on the SIG bar.

The final item in the *setup()* routine is a call to the *checkBatt()* routine to determine the state of the battery pack.

Finally we get to the *loop()* function where the main bulk of the work takes place. The ANDed outputs from the comparators are buffered by a single transistor to generate a target detection signal. If this signal (*compPin*) is low, then a metal target has been detected. As explained earlier, a short delay is introduced to ensure that the sample period aligns with the peak of the target signal from the GB and DISC channel outputs. In these examples, 128 samples are taken and averaged to reduce the effects of noise. The *debugPin* is toggled high at the start of the process, and goes low at the end. Using an oscilloscope it is possible to check the accuracy of the alignment process. This is shown in Fig. 12-1.

The averaged readings are rescaled so that GB ranges from 0 to 511, and DISC ranges from -512 to +511. This is necessary because the GB signal will be positive for all metal targets, whereas the DISC signal can be positive or negative depending on the target material (non-ferrous / ferrous). If the DISC control is advanced from the minimum setting, some higher conductivity items, such as foil and pulltabs, will also end up as a negative signal. The signal strength (SIG) bar is an additional display item for the 4-line LCD.

First the ID display is cleared of any previous reading, and the phase angle associated with the target is calculated by finding the arctangent of the ratio between the DISC and GB signals. This value is in radians, so it needs to be converted to degrees. The VDI number is then simply the integer of the angle in degrees. This is displayed on the LCD.

The magnitude of the target signal (i.e. the signal strength) is calculated by finding the square root of the sum of the squares of the GB and DISC signals. This is also displayed on both the 2-line and 4-line LCDs. In addition, the 4-line LCD displays the signal strength as a moving cursor on the SIG bar.

In addition to the VDI number, a moving cursor is displayed as an alternative indication of the target type. Any previous cursor is erased, taking care not to accidentally remove the VDI bar separator in the middle of the scale.

A custom block character is used as a cursor, and it is positioned according to the current VDI number. Positive numbers will be on the right-hand side, with negative numbers on the left-hand side. This provides a quick indication of whether the metal target is something interesting.

The display counter is then reset, and a delay of 100ms is introduced before the next target signal is analysed.

If no target signal is detected, some other "housekeeping" jobs are carried out. The battery voltage is checked and the battery icons are updated if necessary.

The AM/DISC switch is checked, and an "A" is displayed on the 2-line LCD in the last column of the first line of the LCD for All-Metal mode, or a "D" for Discrimination mode. For the 4-line LCD, the status of the AM/DISC switch is displayed as either "AM" or "DISC".

The display counter is then incremented and, if the count limit is reached, any target information of the display is erased. This happens after approximately 3 seconds.

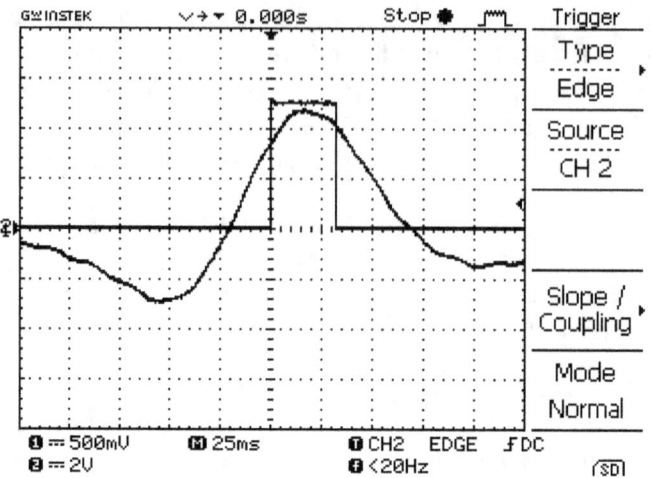

Fig. 12-1: Target Sample Alignment

2-line 16-character LCD

```
// Arduino Nano VLF Metal Detector - Nano #2
// Sketch for 2-line 16-character LCD
// Displays target ID number (VDI), signal strength, target indicator bar, and battery icon
// LCD is cleared after approximately 3 seconds with no target signal

#include <LiquidCrystal.h>

// LCD / Arduino Nano connections
//  1 - GND
//  2 - +5V
//  3 - contrast pot
//  4 - (RS) pin 12
//  5 - (RW) pin 11
//  6 - (EN) pin 10
//  7 - NC
//  8 - NC
//  9 - NC
// 10 - NC
// 11 - (Data4) pin 9
// 12 - (Data5) pin 8
// 13 - (Data6) pin 7
// 14 - (Data7) pin 6
// 15 - (BL+) pin 13 (via resistor)
// 16 - (BL-) GND

// LCD pin assignments (RS, RW, EN, Data7, Data6, Data5, Data4)
LiquidCrystal lcd(12, 11, 10, 9, 8, 7, 6);

// Other pin assignments
byte debugPin = 4;   // For software debugging and scope sync
byte compPin = 5;    // Target trigger from comparator
byte gbPin = A0;     // ADC input for GB channel
byte discPin = A1;   // ADC input for DISC channel
byte battPin = A2;   // ADC input for battery measurement
byte amDisc = 2;     // AM/DISC switch


// Program constants
const byte sampleDelay = 23;        // Number of milliseconds to delay target sample
```

```
const byte numSamples = 128;        // Number of samples for averaging
const word dispCountLimit = 2000;   // Display counter limit

// Program variables
float gb;                  // GB channel reading
float gbAcc = 0;           // Accumulator for GB channel readings
float disc;                // DISC channel reading
float discAcc = 0;         // Accumulator for DISC channel readings
float angleRad;            // Target phase angle in radians
float angleDeg;            // Target angle in degrees
float magnitude;           // Signal magnitude
int vdi = 0;               // Target visual display indicator
int sig = 0;               // Target signal strength
byte vdiBar = 0;           // Position of cursor on VDI bar
float batt;                // Battery voltage reading
word dispCount = 0;        // Clear display counter

void checkBatt() {
  batt = analogRead(battPin);       // Read battery voltage
  batt *= 4.88E-3 * 3;              // Scale ADC reading
  // Display battery condition icons
  lcd.setCursor(15,1);             // Set cursor to column 15, row 1
  if (batt >= 12) {                // Check if battery voltage >= 12V
    lcd.write(byte(0));
  }else{
    if (batt >= 11.7) {            // Check if battery voltage >= 11.7V
      lcd.write(byte(1));
    }else{
      if (batt >= 11.4) {          // Check if battery voltage >= 11.4V
        lcd.write(byte(2));
      }else{
        if (batt >= 11.1) {        // Check if battery voltage >= 11.1V
          lcd.write(byte(3));
        }else{
          if (batt >= 10.8) {      // Check if battery voltage >= 10.8V
            lcd.write(byte(4));
          }else{
            if (batt >= 10.5) {    // Check if battery voltage >= 10.5V
              lcd.write(byte(5));
            }else{
              lcd.print("X");      // Battery needs charging
            }
          }
        }
      }
    }
  }
}

void setup() {
  pinMode(debugPin, OUTPUT);        // Set debug pin as an output
  pinMode(amDisc, INPUT_PULLUP);    // Enable input resistor for AM/DISC switch
  lcd.begin(16,2);                  // Define LCD as having 16 columns and 2 rows
  lcd.clear();                      // Clear LCD
  lcd.setCursor(0,0);               // Set cursor to column 0, row 0
  lcd.print("Arduino Nano VLF");    // Display line 0 title
  lcd.setCursor(0,1);               // Set cursor to column 0, row 1
  lcd.print(" Metal Detector ");    // Display line 1 title
  delay (3000);                     // Display title lines for 3 seconds
  lcd.clear();                      // Clear LCD
  lcd.setCursor(0,0);               // Set cursor to column 0, row 0
  // Set up LCD with ID, SIG markers, and VDI bar
  lcd.print("ID=");
  lcd.setCursor(7,0);
  lcd.print("SIG=");
  lcd.setCursor(0,1);
  lcd.print("------");
  lcd.print("|");
  lcd.print("-------");
```

```
      // Load custom characters into CG RAM of LCD
      // Battery condition icons
      byte batt120V[8] = {14,31,31,31,31,31,31,31};
      lcd.createChar(0, batt120V);
      byte batt117V[8] = {14,31,17,31,31,31,31,31};
      lcd.createChar(1, batt117V);
      byte batt114V[8] = {14,31,17,17,31,31,31,31};
      lcd.createChar(2, batt114V);
      byte batt111V[8] = {14,31,17,17,17,31,31,31};
      lcd.createChar(3, batt111V);
      byte batt108V[8] = {14,31,17,17,17,17,31,31};
      lcd.createChar(4, batt108V);
      byte batt105V[8] = {14,31,17,17,17,17,17,31};
      lcd.createChar(5, batt105V);
      // Block character for VDI bar
      byte block[8] = {31,17,17,17,17,17,31,0};
      lcd.createChar(6, block);
      checkBatt();                    // Check battery voltage
}

void loop() {
      if (digitalRead(compPin) == LOW) {      // Check for target trigger
        delay(sampleDelay);                   // Delay ADC measurements
        gbAcc = 0;                            // Reset GB accumulator
        discAcc = 0;                          // Reset DISC accumulator
        digitalWrite(debugPin, HIGH);         // Set debug pin high
        for (int i=1; i<=numSamples; i++) {   // Take several GB and DISC readings
          gb = analogRead(gbPin);             // Read GB channel
          disc = analogRead(discPin);         // Read DISC channel
          gbAcc += gb;                        // Accumulate GB readings
          discAcc += disc;                    // Accumulate DISC readings
        }
        digitalWrite(debugPin, LOW);          // Set debug pin low
        gb = gbAcc / numSamples;              // Average GB readings
        disc = discAcc / numSamples;          // Average DISC readings
        if (gb >= 512) {                      // Re-scale GB reading
          gb -= 512;
        }
        if (disc >= 512) {                    // Re-scale DISC reading
          disc -= 512;
        } else {
          disc = (512 - disc) * -1;
        }
        lcd.setCursor(3,0);                       // Clear ID display
        lcd.print("    ");
        lcd.setCursor(3,0);
        angleRad = atan(disc/gb);                 // Calculate target phase in radians
        angleDeg = angleRad / TWO_PI * 360;       // Convert to degrees
        vdi = int(angleDeg);                      // Round value for display
        lcd.print(vdi);                           // Display target
        magnitude = sqrt((gb*gb)+(disc*disc));    // Calculate target signal magnitude
        sig = int(magnitude);                     // Round value for display
        lcd.setCursor(11,0);                      // Clear SIG display
        lcd.print("   ");
        lcd.setCursor(11,0);
        lcd.print(sig);                           // Display target signal strength
        if (vdiBar != 6) {                        // Do not overwrite VDI bar separator
          lcd.setCursor(vdiBar,1);                // Set cursor to previous VDI bar position
          lcd.print("-");                         // Overwrite with a "dash"
        }
        vdiBar = vdi / 10 + 6;                    // Calculate position for VDI cursor
        if (vdiBar != 6) {                        // Do not overwrite VDI bar separator
          if (vdiBar > 13) { vdiBar = 13; }       // Do not write beyond end of VDI bar
          lcd.setCursor(vdiBar,1);                // Set cursor to new VDI bar position
          lcd.write(byte(6));                     // Display a block character
        }
        dispCount = 0;                            // Reset display counter
        delay (100);                              // Wait for 100ms before next reading
      }else{
```

```
      checkBatt();                                    // Check battery voltage
      lcd.setCursor(15,0);                            // Set cursor to last character on line 1
      if (digitalRead(amDisc) == LOW){                // Check AM/DISC switch
        lcd.print("D");                               // Detector in DISC mode
      } else {
        lcd.print("A");                               // Detector in AM mode
      }
      dispCount++;                                     // Increment display counter
      if (dispCount >= dispCountLimit) {               // Check if display counter has reached limit
        dispCount = 0;                                 // Reset display counter
        lcd.setCursor(3,0);                            // Clear VDI, SIG, and VDI bar
        lcd.print("      ");
        lcd.setCursor(11,0);
        lcd.print("   ");
        lcd.setCursor(0,1);
        lcd.print("------");
        lcd.print("|");
        lcd.print("-------");
      }
    }
  }
}
```

4-line 16-character LCD

```
#include <LiquidCrystal.h>

// LCD / Arduino Nano connections
//   1 - GND
//   2 - +5V
//   3 - contrast pot
//   4 - (RS) pin 12
//   5 - (RW) pin 11
//   6 - (EN) pin 10
//   7 - NC
//   8 - NC
//   9 - NC
//  10 - NC
//  11 - (Data4) pin 9
//  12 - (Data5) pin 8
//  13 - (Data6) pin 7
//  14 - (Data7) pin 6
//  15 - (BL+) pin 13 (via resistor)
//  16 - (BL-) GND

// LCD pin assignments (RS, RW, EN, Data7, Data6, Data5, Data4)
LiquidCrystal lcd(12, 11, 10, 9, 8, 7, 6);0

// Other pin assignments
byte debugPin = 4;   // For software debugging and scope sync
byte compPin = 5;    // Target trigger from comparator
byte gbPin = A0;     // ADC input for GB channel
byte discPin = A1;   // ADC input for DISC channel
byte battPin = A2;   // ADC input for battery measurement
byte amDisc = 2;     // AM/DISC switch

// Program constants
const byte sampleDelay = 23;      // Number of milliseconds to delay target sample
const byte numSamples = 128;      // Number of samples for averaging
const word dispCountLimit = 2000; // Display counter limit

// Program variables
float gb;            // GB channel reading
float gbAcc = 0;     // Accumulator for GB channel readings
float disc;          // DISC channel reading
float discAcc = 0;   // Accumulator for DISC channel readings
float angleRad;      // Target phase angle in radians
float angleDeg;      // Target angle in degrees
float magnitude;     // Signal magnitude
int vdi = 0;         // Target visual display indicator
int sig = 0;         // Target signal strength
byte vdiBar = 0;     // Position of cursor on VDI bar
byte sigBar = 0;     // Position of cursor on SIG bar
float batt;          // Battery voltage reading
int battV = 0;       // Battery voltage as an integer
int battDisp = 0;    // Battery voltage digit to display
word dispCount = 0;  // Clear display counter

void checkBatt() {
  batt = analogRead(battPin);    // Read battery voltage
  batt *= 4.88E-3 * 3;           // Scale ADC reading
  battV = int(batt * 10);        // Battery voltage to one decimal place
  battDisp = battV / 100 % 10;   // Extract 1st digit
  lcd.setCursor(11,2);
  lcd.print(battDisp);           // Display 1st digit
  battDisp = battV / 10 % 10;    // Extract 2nd digit
  lcd.print(battDisp);           // Display 2nd digit
  lcd.print(".");                // Display decimal place
  battDisp = battV % 10;         // Extract 3rd digit
  lcd.print(battDisp);           // Display 3rd digit
  lcd.print("V");
```

```
}

void setup() {
  pinMode(debugPin, OUTPUT);        // Set debug pin as an output
  pinMode(amDisc, INPUT_PULLUP);    // Enable input resistor for AM/DISC switch
  lcd.begin(16,4);                  // Define LCD as having 16 columns and 4 rows
  lcd.clear();                      // Clear LCD
  lcd.setCursor(0,0);               // Set cursor to column 0, row 0
  lcd.print("Arduino Nano VLF");    // Display line 0 title
  lcd.setCursor(0,1);               // Set cursor to column 0, row 1
  lcd.print(" Metal Detector ");    // Display line 1 title
  delay (3000);                     // Display title lines for 3 seconds
  lcd.clear();                      // Clear LCD
  lcd.setCursor(0,0);               // Set cursor to column 0, row 0
  // Set up LCD with ID, signal strength, VDI bar, and SIG bar
  lcd.print("ID=");
  lcd.setCursor(0,1);
  lcd.print("------");
  lcd.print("|");
  lcd.print("-------");
  lcd.setCursor(0,2);
  lcd.print("SIG=");
  lcd.setCursor(15,2);
  lcd.print("V");
  lcd.setCursor(0,3);
  lcd.print("--------------");
  // Load custom character into CG RAM of LCD
  // Block character for VDI and SIG bars
  byte block[8] = {31,17,17,17,17,17,31,0};
  lcd.createChar(0, block);
  checkBatt();                      // Check battery voltage
}

void loop() {
  if (digitalRead(compPin) == LOW) {        // Check for target trigger
    delay(sampleDelay);                     // Delay ADC measurements
    gbAcc = 0;                              // Reset GB accumulator
    discAcc = 0;                            // Reset DISC accumulator
    digitalWrite(debugPin, HIGH);           // Set debug pin high
    for (int i=1; i<=numSamples; i++) {     // Take several GB and DISC readings
      gb = analogRead(gbPin);               // Read GB channel
      disc = analogRead(discPin);           // Read DISC channel
      gbAcc += gb;                          // Accumulate GB readings
      discAcc += disc;                      // Accumulate DISC readings
    }
    digitalWrite(debugPin, LOW);            // Set debug pin low
    gb = gbAcc / numSamples;                // Average GB readings
    disc = discAcc / numSamples;            // Average DISC readings
    if (gb >= 512) {                        // Re-scale GB reading
      gb -= 512;
    }
    if (disc >= 512) {                      // Re-scale DISC reading
      disc -= 512;
    } else {
      disc = (512 - disc) * -1;
    }
    lcd.setCursor(3,0);                     // Clear ID display
    lcd.print("   ");
    lcd.setCursor(3,0);
    angleRad = atan(disc/gb);               // Calculate target phase in radians
    angleDeg = angleRad / TWO_PI * 360;     // Convert to degrees
    vdi = int(angleDeg);                    // Round value for display
    lcd.print(vdi);                         // Display target
    magnitude = sqrt((gb*gb)+(disc*disc));  // Calculate target signal magnitude
    sig = int(magnitude);                   // Round value for display
    lcd.setCursor(11,0);                    // Clear SIG display
    lcd.print("   ");
    lcd.setCursor(4,2);
    lcd.print(sig);                         // Display target signal strength
```

```
   if (vdiBar != 6) {                          // Do not overwrite VDI bar separator
     lcd.setCursor(vdiBar,1);                  // Set cursor to previous VDI bar position
     lcd.print("-");                           // Overwrite with a "dash"
   }
   vdiBar = vdi / 10 + 6;                       // Calculate position for VDI cursor
   if (vdiBar != 6) {                           // Do not overwrite VDI bar separator
     if (vdiBar > 13) { vdiBar = 13; }          // Do not write beyond end of VDI bar
     lcd.setCursor(vdiBar,1);                   // Set cursor to new VDI bar position
     lcd.write(byte(0));                        // Display a block character
   }
   lcd.setCursor(sigBar,3);                     // Set cursor to previous SIG bar position
   lcd.print("-");                              // Overwrite with a "dash"
   sigBar = sig / 55;                           // Calculate position for SIG bar cursor
   if (sigBar > 13) { sigBar = 13; }            // Do not write beyond end of SIG bar
   lcd.setCursor(sigBar,3);                     // Set cursor to new SIG bar position
   lcd.write(byte(0));                          // Display a block character
   dispCount = 0;                               // Reset display counter
   delay (100);                                 // Wait for 100ms before next reading
 }else{
   lcd.setCursor(11,0);                         // Set cursor to last character on line 1
   if (digitalRead(amDisc) == LOW){             // Check AM/DISC switch
     lcd.print("DISC");                         // Detector in DISC mode
   } else {
     lcd.print("  AM");                         // Detector in AM mode
   }
   dispCount++;                                 // Increment display counter
   if (dispCount >= dispCountLimit) {           // Check if display counter has reached limit
     dispCount = 0;                             // Reset display counter
     lcd.setCursor(3,0);                        // Clear ID, signal strength, VDI and SIG bars
     lcd.print("     ");
     lcd.setCursor(4,2);
     lcd.print("    ");
     lcd.setCursor(0,1);
     lcd.print("------");
     lcd.print("|");
     lcd.print("-------");
     lcd.setCursor(0,3);
     lcd.print("--------------");
     checkBatt();                               // Check battery voltage
   }
 }
}
```

SH1106 OLED 1.3" (128x64) Display

This small display is a 1.3" Zoll OLED from A-Z Delivery. See Fig. 11-4. Note that they made an error when labelling the I2C connections, and named the SCL pin as SCK.

This display uses an SH1106 controller and has 128x64 pixels. To use this display, you need to download the U8g2 library as follows:

In the Arduino IDE, go to Tools > Manage Libraries

Search for U8g2 and select *U8g2 by oliver,* and install.

There are 3 possible modes of operation for this display. The full buffer mode allows graphics to be drawn on the display and is the fastest method, but you can easily run out of RAM with the Arduino Nano. Even if you can get the code to compile and upload to the processor, it may still fail to run. There is also a non-buffered mode but this is very slow. The 3rd mode (used here) is text-only and writes directly to the display. There is no buffer in the micro required, and it can provide up to 8 lines of 16 characters. For this application the text mode works best because of the tiny size of the display.

Organic light emitting diode (OLED) displays are popular with many hobbyists because they can be connected to an Arduino with only a few connections and are not power hungry. OLED displays consist of a matrix of organic LEDs that light up when turned on. Only the pixels that are on use any electricity. There is consequently no backlight required, they have a high contrast, and a wide viewing angle. Plus they are inexpensive when compared to LCDs and TFT displays.

Despite their popularity, many users have problems getting their OLEDs to work because of the difficulty of identifying the correct constructor settings for their display.

The first thing you have to do is to include the library as follows:

#include <U8x8lib.h>

Then configure the constructor:

U8X8_SH1106_128X64_NONAME_HW_I2C u8x8(U8X8_PIN_NONE);

The first part of the constructor specifies that the library name is "U8X8", and the second part declares the display controller as "SH1106_128X64". The display is a generic no-name device, and hence the display name is given as "NO_NAME". The interface we are using is a hardware-based I2C, which only needs two connections (SCL and SDA), plus power and ground. The final part of the constructor instantiates the display as "u8x8", and since this device does not have a reset pin we need to include the "U8X8_PIN_NONE) parameter.

Note that the U8g2 library is only intended for monochrome displays. It is text only, and the only fonts allowed must fit into an 8x8 pixel grid. This mode also uses the least RAM as there is no buffer required in the microcontroller.

In a similar manner to the 4-line LCD, there is a signal strength (SIG) bar as well as one for the VDI. There is only one custom character required, which is the cursor used in the VDI and SIG bars. The LCDs require any custom characters to be loaded into the CG RAM area of the display, but for the OLED the cursor is declared as a variable of type *uint8_t*. This is an unsigned 8-bit character. Unlike (for example) the *int* type, this type will always be 8 characters in size regardless of the processor being used.

```
// Arduino Nano VLF Metal Detector - Nano #2
// Sketch for OLED
// Displays target ID number (VDI), signal strength, and target indicator bar

// Include text-only OLED library
// Writes directly to display
// No buffer in micro required
#include <U8x8lib.h>

// Constructor for AZ Delivery 1,3 Zoll I2C OLED display
U8X8_SH1106_128X64_NONAME_HW_I2C u8x8(U8X8_PIN_NONE);

// OLED / Arduino Nano connections
// VDD - +5V
// GND - GND
// SCK - A5
// SDA - A4

// Pin assignments
byte SDAPin = A4;    // Assign OLED serial data to A4
byte SCLPin = A5;    // Assign OLED serial clock to A5

// Other pin assignments
byte debugPin = 7;   // For software debugging and scope sync
byte compPin = 5;    // Target trigger from comparator
byte gbPin = A0;     // ADC input for GB channel
byte discPin = A1;   // ADC input for DISC channel
byte battPin = A2;   // ADC input for battery measurement
byte amDisc = 2;     // AM/DISC switch

// Program constants
const byte sampleDelay = 23;       // Number of milliseconds to delay target sample
const byte numSamples = 128;       // Number of samples for averaging
const word dispCountLimit = 250;   // Display counter limit

// Program variables
uint8_t block[8] = {0x7e,0x7e,0x7e,0x7e,0x7e,0x7e,0x7e,0x7e};  // Custom block character
float gb;
float gbAcc = 0;
float disc;
float discAcc = 0;
float angleRad;
float angleDeg;
float magnitude;
int vdi = 0;
int sig = 0;
byte vdiBar = 1;
byte sigBar = 1;
float batt;
int battV = 0;
int battDisp = 0;
word dispCount = 0;  // Clear display counter

void checkBatt() {
  batt = analogRead(battPin);   // Read battery voltage
  batt *= 4.88E-3 * 3;          // Scale ADC reading
  battV = int(batt * 10);       // Battery voltage to one decimal place
```

```
    battDisp = battV / 100 % 10;    // Extract 1st digit
    u8x8.setCursor(11,5);
    u8x8.print(battDisp);           // Display 1st digit
    battDisp = battV / 10 % 10;     // Extract 2nd digit
    u8x8.print(battDisp);           // Display 2nd digit
    u8x8.print(".");                // Display decimal place
    battDisp = battV % 10;          // Extract 3rd digit
    u8x8.print(battDisp);           // Display 3rd digit
    u8x8.print("V");
}

void setup()
{
    pinMode(debugPin, OUTPUT);                    // Set debug pin as an output
    pinMode(amDisc, INPUT_PULLUP);                // Enable input resistor for AM/DISC switch
    u8x8.begin();                                 // Initialise display
    u8x8.setFont(u8x8_font_victoriabold8_r);
    u8x8.draw1x2String(0,0,"----------------");   // Display splash screen
    u8x8.draw1x2String(0,2,"ARDUINO NANO VLF");
    u8x8.draw1x2String(0,4," METAL DETECTOR ");
    u8x8.draw1x2String(0,6,"----------------");
    delay(3000);                                  // Display title lines for 3 seconds
    // Set up OLED with ID, SIG markers, and VDI bar
    u8x8.clearDisplay();                          // Clear OLED display
    u8x8.setCursor(0,0);
    u8x8.print("ID=");
    u8x8.setCursor(0,5);
    u8x8.print("SIG=");
    u8x8.setCursor(1,2);
    u8x8.print("------|-------");                 // Display middle marker (vertical line)
    u8x8.setCursor(1,7);
    u8x8.print("--------------");
    checkBatt();
}

void loop() {
    if (digitalRead(compPin) == LOW) {            // Check for target trigger
        delay(sampleDelay);                       // Delay ADC measurements
        gbAcc = 0;                                // Reset GB accumulator
        discAcc = 0;                              // Reset DISC accumulator
        digitalWrite(debugPin, HIGH);             // Set debug pin high
        for (int i=1; i<=numSamples; i++) {       // Take several GB and DISC readings
            gb = analogRead(gbPin);               // Read GB channel
            disc = analogRead(discPin);           // Read DISC channel
            gbAcc += gb;                          // Accumulate GB readings
            discAcc += disc;                      // Accumulate DISC readings
        }
        digitalWrite(debugPin, LOW);              // Set debug pin low
        gb = gbAcc / numSamples;                  // Average GB readings
        disc = discAcc / numSamples;              // Average DISC readings
        if (gb >= 512) {                          // Re-scale GB reading
            gb -= 512;
        }
        if (disc >= 512) {                        // Re-scale DISC reading
            disc -= 512;
        } else {
            disc = (512 - disc) * -1;
        }
        u8x8.setCursor(3,0);                      // Clear ID display
        u8x8.print("      ");
        u8x8.setCursor(3,0);
        angleRad = atan(disc/gb);                 // Calculate target phase in radians
        angleDeg = angleRad / TWO_PI * 360;       // Convert to degrees
        vdi = int(angleDeg);                      // Round value for display
        u8x8.print(vdi);                          // Display target
        magnitude = sqrt((gb*gb)+(disc*disc));    // Calculate target signal magnitude
        sig = int(magnitude);                     // Round value for display
        u8x8.setCursor(4,5);                      // Clear SIG display
        u8x8.print("       ");
```

```
    u8x8.setCursor(4,5);                            // Display target signal strength
    u8x8.print(sig);                                // Do not overwrite VDI bar separator
    if (vdiBar != 7) {                              // Set cursor to previous VDI bar position
      u8x8.setCursor(vdiBar,2);                     // Overwrite with a "dash"
      u8x8.print("-");
    }
    vdiBar = vdi / 10 + 7;                          // Calculate position for VDI cursor
    if (vdiBar != 7) {                              // Do not overwrite VDI bar separator
      if (vdiBar > 14) { vdiBar = 14; };            // Do not write beyond VDI bar position
      u8x8.setCursor(vdiBar, 2);                    // Set cursor to new VDI bar position
      u8x8.drawTile(vdiBar,2,1,block);              // Display a block character
    }
    u8x8.setCursor(sigBar,7);                       // Set cursor to previous SIG bar position
    u8x8.print("-");                                // Overwrite with a "dash"
    sigBar = sig / 55 + 1;                          // Calculate correct position SIG bar indicator
    if (sigBar > 14) { sigBar = 14; }               // Do not write beyond end of SIG bar
    u8x8.setCursor(sigBar,7);                       // Set cursor to new SIG bar position
    u8x8.drawTile(sigBar,7,1,block);                // Display a block character
    dispCount = 0;                                  // Reset display counter
    delay(100);                                     // Wait for 100ms for next target indication
  } else {
    checkBatt();                                    // Check battery voltage
    u8x8.setCursor(12,0);
    if (digitalRead(amDisc) == LOW){                // Check AM/DISC switch
      u8x8.print("DISC");                           // Detector in DISC mode
    } else {
      u8x8.print("AM  ");                           // Detector in AM mode
    }
    dispCount++;                                    // Increment display counter
    if (dispCount >= dispCountLimit) {              // Check if display counter has reached limit
      dispCount = 0;                                // Reset display counter
      u8x8.setCursor(3,0);                          // Clear ID, SIG, and VDI bar
      u8x8.print("      ");
      u8x8.setCursor(4,5);
      u8x8.print("   ");
      u8x8.setCursor(1,2);
      u8x8.print("------|-------");
      u8x8.setCursor(1,7);
      u8x8.print("--------------");
    }
  }
}
```

TFT 2.4" (320x240) Display

This display is a Waveshare 2.4" TFT with a resolution of 320x240 pixels. Communication with the display is via SPI (serial peripheral interface) and is one of the most widely used interfaces between microcontrollers and peripheral ICs.

The SPI pin definitions are:

MISO - (master in slave out) - This is not used by the Waveshare TFT, as this display only needs to read data and does not transmit.

MOSI - (master out slave in) - Note that (due to political correctness) the terms *master* and *slave* are being increasingly phased out in favour of the terms *main* and *subnode*, despite having no connection whatsoever with slavery. The master/slave terminology was first introduced in 1904, and is still frequently used in electronics where one device acts as a master and controls one or more slaves.

CLK - (synchronous clock signal) - The device that generates the clock is the master, and data transmitted between the master and slave nodes is synchronized to the clock. SPI devices can operate at much higher clock speeds compared to the I2C interface used by the OLED.

CS - (chip select) - This line is used by the master to select a slave node. There needs to be a separate chip select line for each slave. The SPI interface supports only one master, but can have one or multiple slaves. In this case there is only one master (the Arduino Nano) and one slave (the TFT display).

DC - (data/command) - This pin tells the display if the data it is receiving is a command or display data.

RESET - (reset) - The name says it all.

LED - (backlight) - This pin is for the backlight control. It can be pulled high (backlight on) or you can apply a pulse-width modulated (PWM) signal at any frequency to control the brightness. It can also be pulled low to turn the backlight off.

The SPI communication protocol is handled by the Adafruit ILI9341 library, which contains all the hardware specific code, and uses the Adafruit GFX library to provide a common syntax and set of graphics functions for a range of display types. The correct timing and data transfer is handled in the background by the Adafruit libraries, so the programmer only has to instantiate the TFT, and then use the library functions to write text and draw graphics on the display.

The software for the TFT is somewhat more complicated than the previous examples. Therefore we will now go through the code step-by-step. As before, the Arduino *sketch* starts with a comment that provides a summary of the operation of the software.

```
// Arduino Nano VLF Metal Detector - Nano #2
// Sketch for TFT display
// Displays target ID number (VDI), signal strength, and confidence bar
```

In order to communicate with the TFT display, we have to include the SPI library, and the Adafruit GFX and ILI9341 libraries. The necessary TFT to SPI mapping is then defined, and the TFT is instantiated as *tft*. This means that all the calls to the Adafruit library function are preceded by *tft*. For example: *tft.print("Hello World")*;

```
// Display is a Waveshare 2.4" TFT (320x240) controlled via Adafruit libraries
#include <SPI.h>
#include <Adafruit_GFX.h>
#include <Adafruit_ILI9341.h>

// TFT SPI definitions
#define TFT_MISO 12  // Master in Slave out (MISO) - not used by Waveshare TFT
#define TFT_MOSI 11  // Master out Slave in (MOSI)
#define TFT_SCK 10   // Synchronous clock signal
#define TFT_CS 9     // Chip select
#define TFT_DC 8     // Data / Control
#define TFT_RESET 7  // Reset
#define TFT_LED 6    // Backlight

// Instantiate ILI9341 TFT
Adafruit_ILI9341 tft = Adafruit_ILI9341(TFT_CS, TFT_DC, TFT_MOSI, TFT_SCK, TFT_RESET, TFT_MISO);
```

Specific variables are then assigned to various microcontroller pins. Note that a debug pin is also assigned to assist with software debugging, and to time certain events in conjunction with an oscilloscope.

The comparator output is used to trigger analysis of the GB and DISC channel outputs and subsequently provide a target ID, VDI bar, signal (DEPTH) indication, and a confidence bar.

Remaining assignments are for the battery voltage and the AM/DISC switch.

```
// Other pin assignments
byte debugPin = 4;   // For software debugging and scope sync
byte compPin = 5;    // Target trigger from comparator
byte gbPin = A0;     // ADC input for GB channel
byte discPin = A1;   // ADC input for DISC channel
byte battPin = A2;   // ADC input for battery measurement
byte amDisc = 2;     // AM/DISC switch
```

Next we need to declare some useful program constants.

The *sampleDelay* constant (as described earlier) is used to introduce a delay between when the target trigger is initiated and the point where the Nano samples the signal at the outputs of the GB and DISC channels. With the correct delay, the *debug* pulse should be aligned with the positive peak of the double-differentiated channel outputs. This is also important for obtaining accurate VDI numbers.

The display counter limit of 200 provides an approximate time of 3 seconds for the target information to remain on the display before being automatically erased.

```
// Program constants
const byte sampleDelay = 23;        // Number of milliseconds to delay target sample
const byte numSamples = 128;        // Number of samples for averaging
const word dispCountLimit = 200;    // Display counter limit
```

The purpose of the declared variables should be fairly obvious once you read the comments. The trigonometrical arctangent function performs its calculations in radians, which is why the code includes variables for both radians and degrees. For display purposes we need the phase angle in degrees. The purpose of some of the variables will become clearer as we go through the code.

```
//Program variables
float gb;                                         // GB channel reading
float gbAcc = 0;                                  // Accumulator for GB channel readings
float disc;                                       // DISC channel reading
float discAcc = 0;                                // Accumulator for DISC channel readings
float angleRad;                                   // Target phase in radians
float angleDeg;                                   // Target phase in degrees
float magnitude;                                  // Signal magnitude
int vdi = 0;                                       // Target visual display indicator
byte vdiBar = 1;                                   // Position of cursor on VDI bar
int sig = 0;                                        // Signal strength
int sigBar;                                         // Length of signal bar
int prevSigBar = 0;                                 // Previous length of signal bar
int sigBarY;                                        // Y position of cursor on SIG bar
int prevSigBarY = 194;                              // Previous Y position of cursor on SIG bar
int temp;                                           // Temporary variable
byte mode = 0;                                      // Operating mode (AM or DISC)
byte prevMode = 2;                                  // Previous operating mode
float batt;                                         // Battery voltage reading
int battV = 0;                                      // Battery voltage as an integer
int battDisp = 0;                                   // Battery voltage digit to display
word dispCount = 0;                                 // Clear display counter
byte backlight = 192;                               // PWM value for backlight brightness
byte targetID;                                      // Target ID number
uint16_t targetIDColor;                             // Target ID colour (green or red)
uint16_t battColor;                                 // Battery voltage display colour (green or red)
uint16_t prevBattColor = ILI9341_BLACK;             // Previous battery voltage colour
byte battArray[] = {99, 99, 99, 99, 99};            // Record of battery voltage digits
byte acceptBar = 0;                                 // Number of accept bars for CONF display
byte rejectBar = 0;                                 // Number of reject bars for CONF display
```

A special function is included for measuring the battery voltage. The Nano's ADC has a resolution of 4.88mV, and a potential divider reduces the battery voltage to bring it within the range of the ADC. Hence we have to multiply the ADC reading by 4.88E-3 * 3 to get the actual battery voltage. In addition, the voltage is displayed as a green colour if the voltage is greater or equal to 10.5V, and red if it is lower. The battery voltage is provided in the lower section of the display, as shown in Fig. 13-35, and the individual digits are only updated if the voltage or its display changes colour. The reason for checking the colour change is because the decimal point and the voltage sign also need to change colour, otherwise the voltage display could become a mixture of red and green characters. Unlike the LCD and OLED displays, the TFT does not write the complete font character over any previous character, but only lights the appropriate pixels, which means that you can end up with an unreadable mess on the screen. The simple solution is to first overwrite the existing character in the background colour. In this case, the background colour is black.

```
void checkBatt() {
  batt = analogRead(battPin);     // Read battery voltage
  batt *= 4.88E-3 * 3;            // Scale ADC reading
  battV = int(batt * 10);         // Battery voltage to one decimal place
  if (battV >= 105) {
    battColor = ILI9341_GREEN;    // Display colour is green when voltage >= 10.5V
  } else {
    battColor = ILI9341_RED;      // Display colour is red when battery voltage < 10.5V
  }
  // Only update battery voltage display digits if voltage has changed
  battDisp = battV / 100 % 10;    // Extract 1st digit
  if ((battArray[0] != battDisp) | (battColor != prevBattColor)) {
    tft.drawChar(128, 214, 32, battColor, ILI9341_BLACK, 2);   // Overwrite previous value
    tft.drawChar(128, 214, battDisp + 48, battColor, ILI9341_BLACK, 2);  // Display 1st digit
    battArray[0] = battDisp;      // Save new value for 1st digit
  }
  battDisp = battV / 10 % 10;     // Extract 2nd digit
  if ((battArray[1] != battDisp) | (battColor != prevBattColor)) {
    tft.drawChar(140, 214, 32, battColor, ILI9341_BLACK, 2);   // Overwrite previous value
    tft.drawChar(140, 214, battDisp + 48, battColor, ILI9341_BLACK, 2);  // Display 2nd digit
    battArray[1] = battDisp;      // Save new value for 2nd digit
  }
  if ((battArray[2] != 46) | (battColor != prevBattColor)) {
    tft.drawChar(152, 214, 32, battColor, ILI9341_BLACK, 2);
    tft.drawChar(152, 214, 46, battColor, ILI9341_BLACK, 2);   // Display decimal place
    battArray[2] = 46;
  }
  battDisp = battV % 10;    // Extract 3rd digit
  if ((battArray[3] != battDisp) | (battColor != prevBattColor)) {
    tft.drawChar(164, 214, 32, battColor, ILI9341_BLACK, 2);   // Overwrite previous value
    tft.drawChar(164, 214, battDisp + 48, battColor, ILI9341_BLACK, 2);  // Display 3rd digit
    battArray[3] = battDisp;      // Save new value for 3rd digit
  }
  if ((battArray[4] != 86) | (battColor != prevBattColor)) {
    tft.drawChar(176, 214, 32, battColor, ILI9341_BLACK, 2);
    tft.drawChar(176, 214, 86, battColor, ILI9341_BLACK, 2);   // Display voltage sign
    battArray[4] = 86;
  }
  prevBattColor = battColor;   // Save battery voltage colour
}
```

The *setup()* routine is always executed first following power up or reset. Initially the pin modes are defined. In this case we only need to define the *debugPin* as an output. The AM/DISC switch input is analog, and does not need to be defined since this is the default mode. However, we do need to specify that the input requires the internal 20k pullup resistor to be enabled. When the AM/DISC switch is on, the input is pulled low.

Next the static (unchanging) parts of the display are created, such as the screen border, inner display area, VDI bar, signal strength (DEPTH) bar, and confidence (CONF) bar.

The final item in the *setup()* routine is a call to the *checkBatt()* routine to determine the state of the battery pack.

Arduino Nano VLF Metal Detector

```
void setup()
{
  pinMode(debugPin, OUTPUT);       // Set debug pin as an output
  pinMode(amDisc, INPUT_PULLUP);   // Enable pullup resistor for AM/DISC switch
  analogWrite(TFT_LED, backlight); // Turn on TFT backlight
  tft.begin();                     // Init display
  tft.setRotation(3);              // Rotate display to landscape with connector on RHS
  tft.fillScreen(ILI9341_BLACK);   // Clear display

  // Draw screen border
  tft.drawRect(0, 0, 320, 240, ILI9341_WHITE);

  // Inner display area
  tft.drawLine(98, 239, 98, 142, ILI9341_WHITE);
  tft.drawLine(98, 142, 88, 132, ILI9341_WHITE);
  tft.drawLine(88, 132, 88, 48, ILI9341_WHITE);
  tft.drawLine(88, 48, 98, 38, ILI9341_WHITE);
  tft.drawLine(98, 38, 222, 38, ILI9341_WHITE);
  tft.drawLine(222, 38, 232, 48, ILI9341_WHITE);
  tft.drawLine(232, 48, 232, 132, ILI9341_WHITE);
  tft.drawLine(232, 132, 222, 142, ILI9341_WHITE);
  tft.drawLine(222, 142, 222, 239, ILI9341_WHITE);
  tft.drawLine(98, 142, 222, 142, ILI9341_WHITE);
  tft.drawLine(98, 170, 222, 170, ILI9341_WHITE);
  tft.drawLine(98, 204, 222, 204, ILI9341_WHITE);

  // Target ID text
  tft.setCursor(107, 150);
  tft.setTextColor(ILI9341_WHITE);
  tft.setTextSize(2);
  tft.print("TARGET ID");

  // Draw VDI bar
  tft.drawRect(61, 10, 195, 18, ILI9341_WHITE);

  // Draw signal (DEPTH) bar
  tft.drawRect(30, 48, 22, 118, ILI9341_WHITE);

  // Depth text
  tft.setCursor(12, 182);
  tft.setTextColor(ILI9341_WHITE);
  tft.setTextSize(2);
  tft.print("DEPTH");

  // Draw confidence (CONF) bar
  tft.drawRect(267, 48, 22, 118, ILI9341_WHITE);

  // Confidence bar text
  tft.setCursor(254, 182);
  tft.setTextColor(ILI9341_WHITE);
  tft.setTextSize(2);
  tft.print("CONF");

  checkBatt();  // Check battery voltage
}
```

Finally we get to the *loop()* function where the main bulk of the work takes place. The ANDed outputs from the comparators are buffered by a single transistor to generate a target detection signal. If this signal (*compPin*) is low, then a metal target has been detected. As explained earlier, a short delay is introduced to ensure that the sample period aligns with the peak of the target signal from the GB and DISC channel outputs. In this example, 128 samples are taken and averaged to reduce the effects of noise. The *debugPin* is toggled high at the start of the process, and goes low at the end. Using an oscilloscope it is possible to check the accuracy of the alignment process. This is shown in Fig. 12-1.

The averaged readings are rescaled so that GB ranges from 0 to 511, and DISC ranges from -512 to +511. This is necessary because the GB signal will be positive for all metal targets, whereas the DISC signal can be positive or negative depending on the target material (non-ferrous / ferrous). If the DISC control is advanced from the minimum setting, some lower conductivity items, such as foil and pulltabs, will also end up as a negative signal.

```
void loop() {
  if (digitalRead(compPin) == LOW) {      // Check for target trigger
    delay(sampleDelay);                    // Delay ADC measurements
    gbAcc = 0;                             // Reset GB accumulator
    discAcc = 0;                           // Reset DISC accumulator
    digitalWrite(debugPin, HIGH);          // Set debug pin high
    for (int i = 1; i <= numSamples; i++) {  // Take a number of GB and DISC readings
      gb = analogRead(gbPin);              // Read GB channel
      disc = analogRead(discPin);          // Read DISC channel
      gbAcc += gb;                         // Accumulate GB readings
      discAcc += disc;                     // Accumulate DISC readings
    }
    digitalWrite(debugPin, LOW);           // Set debug pin low
    gb = gbAcc / numSamples;               // Average GB readings
    disc = discAcc / numSamples;           // Average DISC readings
    if (gb >= 512) {                       // Re-scale GB reading
      gb -= 512;
    }
    if (disc >= 512) {                     // Re-scale DISC reading
      disc -= 512;
    } else {
      disc = (512 - disc) * -1;
    }
```

The phase angle introduced by the target is then determined by calculating the arctangent of the ratio between the DISC and GB readings. The result is in radians, and needs to be converted to degrees before being displayed as a target ID.

```
    angleRad = atan(disc/gb);              // Calculate target phase in radians
    angleDeg = angleRad / TWO_PI * 360;    // Convert to degrees
    vdi = int(angleDeg);                   // Round value for display
```

At the top of the display is a VDI bar which shows a simple indication of the target phase. If the target is non-ferrous, the cursor is displayed in a green colour on the right-hand side of the display. Alternatively, if the target is ferrous, the cursor is displayed in red on the left-hand side. By adjusting the DISC control clockwise, certain lower conductivity targets such as foil and pulltabs may also be shown in red. In this case it is strictly more accurate to say that accepted targets are on the right, and rejected targets are on the left. When the detector is operating in DISC mode, the rejected targets may only appear if they break through as chatter, such as when a large ferrous object is close to the coil

Unlike the LCD and OLED displays, there is no requirement to create any custom characters. For the TFT display we can simply use the appropriate ASCII character as a cursor. But ... if you remember our discussion in Chapter 11 (Fig. 11-6) concerning the CP437 character set and the error in the Arduino's built-in font, you will recall that the missing character is decimal 176. The result is that every following character is off by one. According to the standard ASCII character set, the character we require is decimal 219, but due to the Arduino font error we must specify 218.

Note that we must also erase any existing cursor, before drawing the new one, by overwriting it in black.

```
// Overwrite previous VDI bar cursor
tft.drawChar(vdiBar, 11, 32, ILI9341_GREEN, ILI9341_BLACK, 2);
vdiBar = vdi + 90 + 63;   // Calculate new VDI bar position
if (vdi >= 0) {
    // Display cursor in green when positive
    tft.drawChar(vdiBar, 11, 218, ILI9341_GREEN, ILI9341_BLACK, 2);
} else {
    // Display cursor in red when negative
    tft.drawChar(vdiBar, 11, 218, ILI9341_RED, ILI9341_BLACK, 2);
}
```

Below the VDI bar is where the target ID is displayed. The target ID can range from -90 to + 90 degrees, and any ID above (and including) zero is shown in green. Any ID less than zero is shown in red. The minus sign is not displayed.

The target ID is split into two separate digits using the Arduino div and mod commands, and displayed using a large font size.

```
if (vdi >= 0) {
    targetIDColor = ILI9341_GREEN;   // Display target ID in green when positive
} else {
    targetIDColor = ILI9341_RED;     // Display target ID in red when negative
    vdi = vdi * -1;                  // Remove negative sign
}
targetID = vdi / 10 % 10 + 48;       // Calculate new value for target ID
// Display first digit of target ID
tft.drawChar(110, 55, targetID, targetIDColor, ILI9341_BLACK, 8);
targetID = vdi % 10 + 48;
// Display second digit for target ID
tft.drawChar(170, 55, targetID, targetIDColor, ILI9341_BLACK, 8);
```

On the right-hand side of the display is the confidence (CONF) bar. This reacts differently to other confidence bars such as the one provided on the Fisher F75. On the Arduino Nano VLF detector, the bar moves upwards from the centre whenever an accepted target is detected, and moves downwards for every rejected target. When above the centre the bar is displayed in the colour green, and in red when below the centre. This feature is very useful when trying to identify if a non-ferrous target is hiding amongst ferrous trash. By moving the coil in small sweeps from side to side a comparison of accepted to rejected targets can be obtained, and assists in determining whether to dig or not.

```
if (targetIDColor == ILI9341_GREEN) {       // Check if target ID is positive
  if (acceptBar >= rejectBar) {              // Check if accept bar is already active
    if (acceptBar != 8) { acceptBar++; }     // Limit extent of accept bar
    // Increase accept bar
    tft.fillRect(269, 106 - acceptBar * 7, 18, acceptBar * 7, ILI9341_GREEN);
  } else {
    // Limit extent of reject bar
    if (rejectBar != 0) { rejectBar--; }
    // Decrease reject bar
    tft.fillRect(269, 108 + rejectBar * 7, 18, (8 - rejectBar) * 7, ILI9341_BLACK);
  }
} else {
  if (rejectBar >= acceptBar) {              // Check if reject bar is already active
    if (rejectBar != 8) { rejectBar++; }     // Limit extent of reject bar
    tft.fillRect(269, 108, 18, rejectBar * 7, ILI9341_RED);   // Increase reject bar
  } else {
    if (acceptBar != 0) { acceptBar--; }     // Limit extent of accept bar
    // Decrease accept bar
    tft.fillRect(269, 50, 18, (8 - acceptBar) * 7, ILI9341_BLACK);
  }
}
```

The magnitude of the target signal is determined by calculating the square root of the sum of the squares of the DISC and GB channel outputs. Although the relationship between the target depth and the signal strength is not strictly speaking logarithmic, it's close enough for our purposes and helps to linearise the movement of the signal bar. Many people get caught out by the Arduino log functions because *log()* is the natural logarithm to the base e, whereas *log10()* is the logarithm to the base 10. We need to use the latter.

As with other parts of the display, the signal bar only updates if the signal changes from the previous value.

```
magnitude = sqrt((gb*gb)+(disc*disc));   // Calculate target signal magnitude
sig = int(magnitude);                    // Round value for display
sig -= 22;                               // Offset signal value
if (sig <= 0) { sig = 1; }               // Limit extent of signal value
sigBar = int(log10(sig) * 10) * 4;       // Find logarithm of signal and scale result
sigBarY = 163 - sigBar;                  // Offset start of signal bar
if (sigBar >= prevSigBar) {              // Only update signal bar if necessary
  // Increase signal bar
  tft.fillRect(32, sigBarY, 18, sigBar, ILI9341_WHITE);
} else {
  // Decrease signal bar
  tft.fillRect(32, prevSigBarY, 18, prevSigBar - sigBar, ILI9341_BLACK);
}
prevSigBar = sigBar;                     // Save length of signal bar
prevSigBarY = sigBarY;                   // Save Y position of signal bar
```

Before going around the loop again, the display count is reset, and a delay of 100ms is introduced between VDI target indications.

```
    dispCount = 0;                              // Reset display count
    delay(100);                                 // Wait for 100ms between VDI target indications
```

If no target signal is detected the software performs some other housekeeping tasks.

Firstly the AM/DISC switch is checked and, if the mode has been changed, the current mode is updated on the display above the battery voltage.

```
  } else {
    tft.setTextSize(2);
    mode = digitalRead(amDisc);                 // Check AM/DISC switch
    if (mode != prevMode) {                     // Only update display if necessary
      if (mode == LOW) {                        // Test if DISC mode selected
        tft.setCursor(135, 180);
        tft.setTextColor(ILI9341_BLACK);
        tft.print(" AM ");                      // Overwrite AM text
        tft.setCursor(135, 180);
        tft.setTextColor(ILI9341_YELLOW);
        tft.print("DISC");                      // Display mode as DISC
      } else {                                  // Otherwise mode must be AM
        tft.setCursor(135, 180);
        tft.setTextColor(ILI9341_BLACK);
        tft.print("DISC");                      // Overwrite DISC text
        tft.setCursor(135, 180);
        tft.setTextColor(ILI9341_YELLOW);
        tft.print(" AM ");                      // Display mode as AM
      }
    }
    prevMode = mode;                            // Save operating mode
```

Secondly the display counter is incremented. There is also a small but important difference here when compared to the LCD and OLED examples. Note there is a delay of 10ms added to the loop. This is because interaction with the TFT is faster than with the other displays, and display updating only occurs if something has changed. Basically, it gets around the loop too quickly and we need to slow it down.

```
    dispCount++;                                // Increment display counter
    delay(10);                                  // Add a 10ms delay to loop
```

If the display counter reaches the limit, which is after approximately 3 seconds, the display counter is reset. Then the battery voltage is checked and updated if necessary. Finally the target ID, VDI bar, signal (DEPTH) bar, and confidence (CONF) bar are all cleared.

```
    if (dispCount >= dispCountLimit) {      // Check if display count has reached limit
      dispCount = 0;                        // Clear display count
      checkBatt();                          // Check battery voltage
      tft.drawChar(110, 55, 32, ILI9341_WHITE, ILI9341_BLACK, 8);  // Clear target ID
      tft.drawChar(170, 55, 32, ILI9341_WHITE, ILI9341_BLACK, 8);
```

```
      tft.fillRect(64, 11, 191, 16, ILI9341_BLACK);      // Clear VDI bar
      tft.setCursor(30, 207);
      tft.setTextColor(ILI9341_BLACK);
      tft.setTextSize(2);
      tft.fillRect(32, 50, 18, 114, ILI9341_BLACK);      // Clear signal bar
      tft.fillRect(269, 50, 18, 114, ILI9341_BLACK);     // Clear confidence bar
      acceptBar = 0;                                     // Save length of accept bar
      rejectBar = 0;                                     // Save length of reject bar
    }
  }
}
```

Chapter 13

Building the Detector

"It does not matter how slowly you go as long as you do not stop."

--- Confucius

Please read:

If you go through the following steps carefully, and in order, you will have a greater chance of ending up with a detector that actually works. Please do <u>not</u> proceed to the next step until you have solved any problems encountered in the current step. If you get stuck at any stage during the build, then seek help in the Geotech forums before moving on. Do <u>not</u> simply populate the whole board; discover it doesn't work (which will be the most likely result) and then post in the forums: "I've built the Arduino Nano VLF Detector and it's not working. What's wrong?" .... as you will receive little sympathy from the Geotech members.

To reiterate: **Please follow the instructions step-by-step.**

Important Points:

Check the value of each component before you start soldering. Fitting an incorrect value in the wrong place is one of the most common mistakes.

The negative terminal of the battery is treated as ground (0V) in this design. All voltages are referenced to this point.

Please refer to the component parts list (BOM) in Appendix A, and the full set of schematics shown in Appendix B.

Step 1:   **Fit all IC sockets (Fig. 13-1)**

For A1 and A2 use 2x 15-pin female headers for each Arduino Nano.

U2 (LT1054) ... 8-pin

U3 (LT1054) ... 8-pin

U4 (TL072) ... 8-pin

U5 (CD14538BE) ... 16-pin

U6 (CD14538BE) ... 16-pin

U7 (4066) ... 14-pin

U8 (LM393) ... 8-pin

U9 (TL071) ... 8-pin

U10 (TL072) ... 8-pin

U11 (TL072) ... 8-pin

Fig. 13-1: IC Sockets

Step 2:   Fit all connectors (see Fig. 13-2)

J1 – Battery (2-pin)

J2 – Ground balance pot (2-pin)

J3 – TX coil (2-pin)

J4 – DISC pot (2-pin)

J5 – Sensitivity pot (3-pin)

J6 – RX coil (2-pin)

J7 – AM / DISC switch-A (2-pin)

J8 – RX tuning capacitor (2-pin)

J9 – Loudspeaker / headphones (2-pin)

J10 – Unassigned ports (6-pin)

J11 – LCD / TFT display (9-pin)

J12 – AM / DISC switch-B (2-pin)

J13 – OLED (4-pins)

Note that J8 is a screw-type connector (middle top edge of the PCB – see Fig. 13-2) to enable the RX tuning capacitor to easily be swapped during experimentation. If such flexibility is not required, the RX tuning capacitor may be fitted directly in place of the connector.

Step 3:   Build voltage converters

Fit R17 (20k), R18 (10k), C2, C3, C5, C6 (10μF), C1, C4, C7 (470μF), Q1 (2N3904), TP1, TP2, TP3, TP4 (test points), U1 (79L05), U2 (LT1054) and U3 (LT1054).

Make sure that you insert the electrolytic capacitors with the correct polarity. The PCB is laid out so that the negative terminal of the polarized capacitors are all at the bottom. Note that you will not be able to check the power supply voltages until Step 6, as the Arduino Nano must first be installed and programmed.

The Arduino Nano programming procedure is described in Step 5, and the voltage converter circuitry is shown in Fig. 13-3.

Fig. 13-2: Connectors

Fig. 13-3: Voltage Converters

Step 4:   Arduino Nano #1

Fit A1 (Arduino Nano).

Note correct orientation for the device with the USB connector at the lower edge of the PCB.

Fig. 13-4: Arduino Nano #1

Step 5:   Program the Arduino Nano

Download the Arduino Integrated Development Environment (IDE). Note that some compatible Nano boards use a CH340 chip instead of the original FT232 USB UART IC. In programming terms these boards are 100% compatible, as are the other hardware functions. The manufacturer may tell you to download a driver from their site to allow the bootloader to be used. In my particular case there was no driver to be found on the manufacturer's website, and I eventually discovered that it was already included with the Linux Mint operating system.

Be aware that a lot of cheap Mini-USB cables only contain wiring for power, and the data lines are missing. These cables can only be used for charging, and not for communications. If you have a problem connecting to the Arduino Nano, then it's worth checking the cable before you suspect it's a driver issue.

Download the Arduino Nano #1 sketch from the Geotech website, and connect the Nano to your computer using a Mini-B USB data cable.

Launch the Arduino IDE and select the serial device of the Nano board using: Tools > Port. In my particular case (using Linux Mint) this was "/dev/tty/USB0".

Upload the sketch to the Nano board, and the TX and RX LEDs on the board will flash. If successful, a message "Done uploading" will appear in the status bar.

Step 6:   Test power supplies

Attach a 12V battery supply to J1, making absolutely certain that you've connected the battery with the correct polarity. Don't falsely believe that this could never happen to you, because (trust me) it definitely can. Just remember, that where electronics is concerned, Murphy is always watching.

The Arduino Nano is used to provide the +5V supply to the analog circuitry. If you want to double-check the +5V supply then connect a voltmeter between A1 pin 27 and TP1.

The other power supply voltages can be checked at the following test points with reference to TP1 (0V).

TP2 = -12V

TP3 = -12V

TP4 =   -5V

Step 7:   R-2R ladder network

Fit R1, R2, R4, R6, R8, R10, R12, R14, R16 (20k), and R3, R5, R7, R9, R11, R13, R15 (10k).

Attach a scope probe to the junction of R15 / R16, as shown in Fig. 13-5. You should observe a sine wave similar to that in Fig. 13-6.

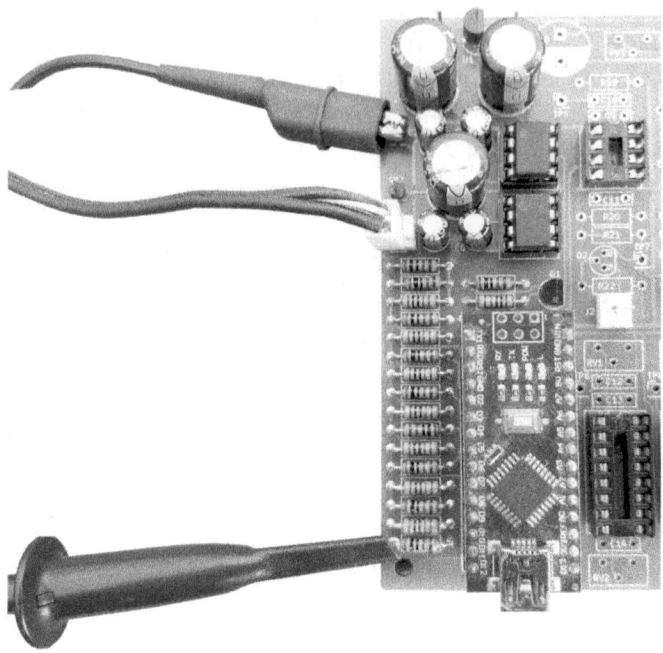

Fig. 13-5: R-2R Ladder Scope Connection

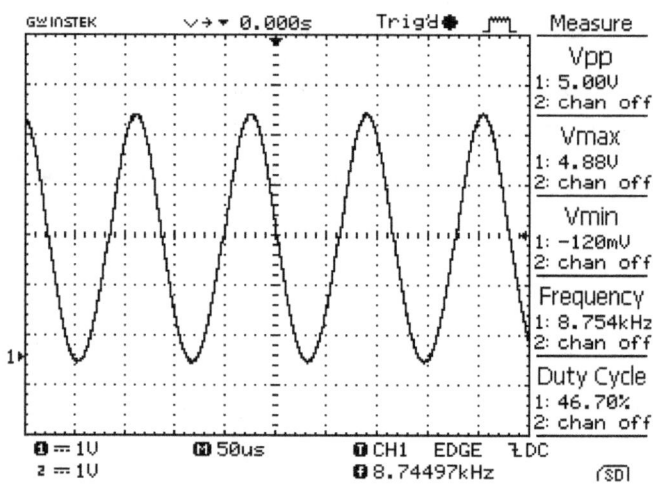

Fig. 13-6: R-2R Ladder Output

Step 8:   **Transmitter Circuit**

Fit R19, R20 (10k), R23, R24 (10R), R25 (18k), RV3 (10k preset), C9, C11 (100nF), C10 (1µF), C8 (470µF), TP5, TP9 (test point), D1, D2 (1N4148), Q4 (2N3904), Q5 (2N3906), and U4 (TL072).

The TX circuitry is shown in Fig. 13-7.

Measure the buffered R-2R ladder output at TP5 (which should be the same as Fig. 13-6), and then the TX output at TP9 (see Fig. 13-8). Note that it is not necessary to attach a coil at this point since the TX circuit is a forced oscillator.

With reference to Fig. 13-8 and Fig. 13-9, adjust the TX bias setting using RV3 (with the power supply input set to 10V) to remove any clipping that may be present on the negative half of the TX output.

Fig. 13-7: Transmitter (TX)

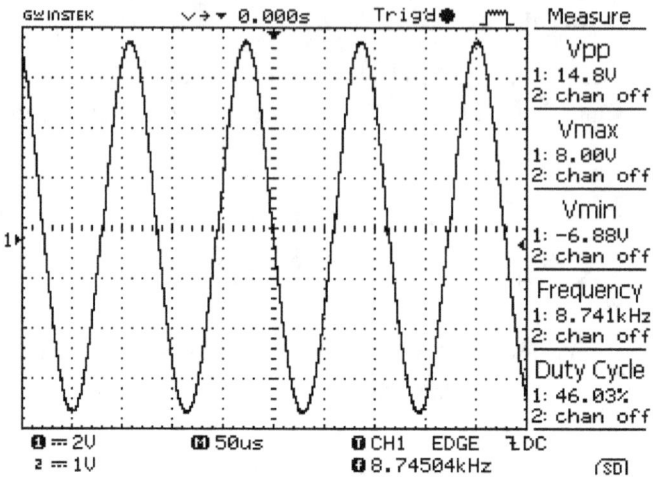

Fig. 13-8: TX Output (No Clipping)

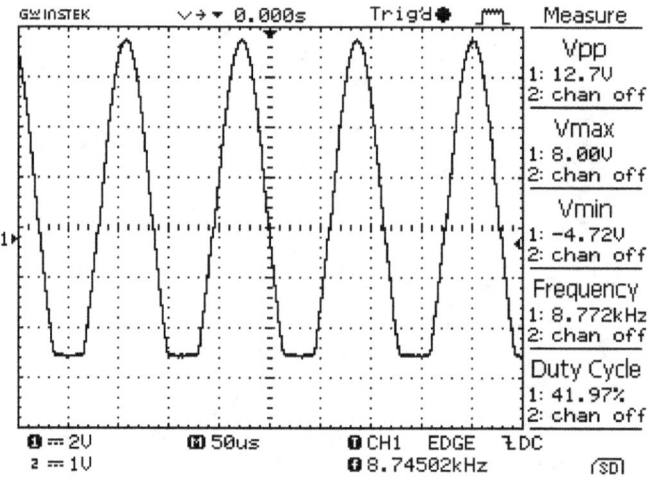

Fig. 13-9: TX Output (Clipping)

Step 9:   Preamp

Fit R28, R30 (5k1), R29, R31 (100k), R36 (1k), C23, C24 (100nF), C18 (220nF), TP11 (test point), and U9 (TL071).

Before testing the preamp it will be necessary to connect a search coil to the TX (J3) and RX (J6) connectors, and a tuning capacitor(s) to J8.

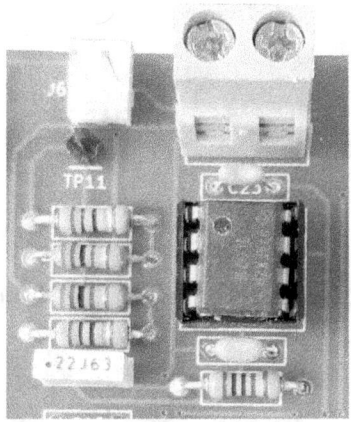

Fig. 13-10: Preamp

Step 10: Search coil

A Garrett ACE concentric search coil (9" x 6.5" elliptical) was used for the initial setup and testing. Coils from other manufacturers will be investigated in Chapter 14, although the procedure that follows is quite similar.

Since the transmitter frequency is controlled by the Arduino Nano software, the TX coil inductance is unimportant. However, the RX inductance will be part of an LC tuned circuit, and we need to decide on a suitable capacitor value to connect to J8.

From Figures 13-6 and 13-8 it can be seen that the TX frequency is close to 8.75kHz. If we offset the RX resonant frequency to be lower than the TX at around 8kHz, this will reduce susceptibility to phase noise on the TX or modulation due to metal targets. With reference to Fig. 13-11, note that the RX coil has an inductance of 1.7mH.

Then calculating the value of the tuning capacitor:

$$C = \frac{1}{\omega^2 L} = \frac{1}{(2\pi \times 8k)^2 \times 1.7m} = 232.8nF$$

Using standard values, a 220nF capacitor in parallel with the RX coil will result in an actual resonant frequency of:

$$f = \frac{1}{2\pi\sqrt{LC}} = \frac{1}{2\pi\sqrt{1.7m \times 220n}} = 8.23kHz$$

As mentioned in Chapter 5, and from Fig. 13-11, it is important to know which side of the TX and RX coils are referenced to 0V when you wire them to connectors J3 and J6, otherwise strange things can happen. The same applies to search heads where the TX and RX coils share a common connection.

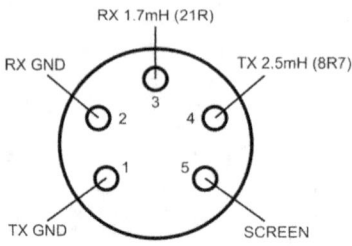

Fig. 13-11: Original Garrett ACE Coil Connections

Note that the Garrett ACE coils are unusual in that the inductance of the RX coil is lower than the TX inductance. Pin 5 is connected to the screen of the cable, which is in turn connected to the coil's electrostatic shield. There is no electrical connection between pins 1, 2 and 5, which means you will need to connect pin 5 to 0V on the PCB. The simplest way to achieve this is to connect pins 1 and 5 together on the back of the coil socket.

If you plan to wire the coil connector to match the Garrett ACE coil, then the pin mapping to the TX (J3) and RX (J6) connectors is shown in Fig. 13-12.

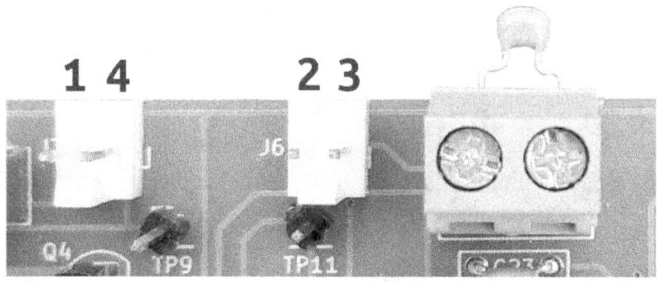

Fig. 13-12: Coil Connections for Garrett ACE (J3 and J6)

Step 11: Connect the search coil

With reference to Figures 13-11 and 13-12, it should be clear that the left-hand pins of J3 and J6 will be electrically connected to 0V on the PCB. Attach channel 1 scope probe to TP9 (TX output signal) and channel 2 probe to TP11 (preamp output). Trigger the scope from channel 1. You should then see the signals shown in Fig. 13-13 with no target present, noting that there is a phase shift between the TX and RX signals. The lower amplitude signal is the RX.

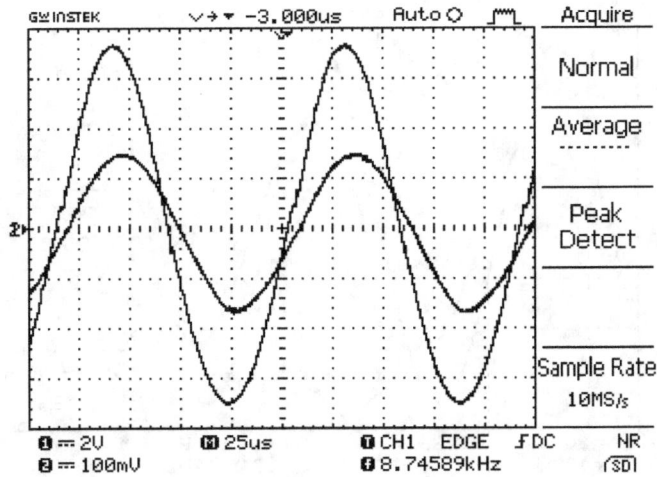

Fig. 13-13: TX and RX Signals (Garrett ACE Coil) with No Target Present

Now place a non-ferrous target (such as a coin) near the centre of the coil, and note that the RX signal shifts right and decreases in amplitude, as shown in Fig. 13-14.

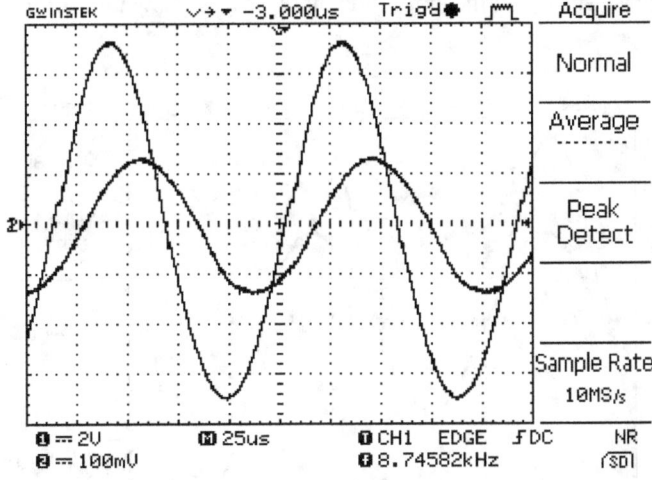

Fig. 13-14: TX and RX Signals (Garrett ACE Coil) with a Non-ferrous Target

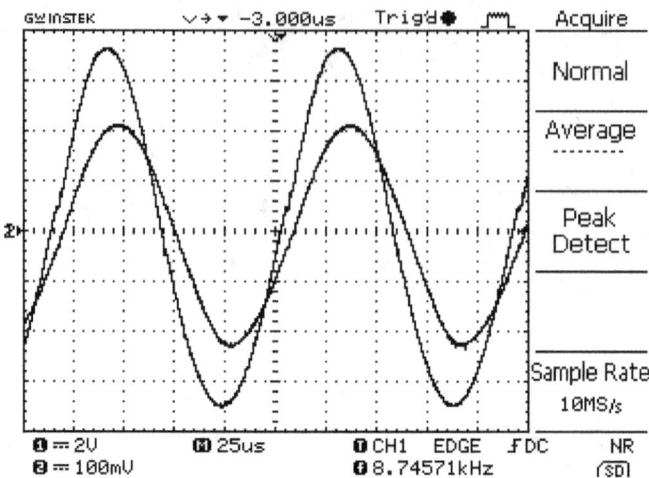

Fig. 13-15: TX and RX Signals (Garrett ACE Coil) with Ferrous Target

Replace the non-ferrous item with a ferrous target (such as a nail), and note that the RX signal has a negligible phase shift, but this time increases in amplitude. This is shown in Fig. 13-15.

Finally, replace the ferrous item with a ferrite core to simulate ground response. In this case the RX signal exhibits no additional phase shift, and the amplitude increases, as shown in Fig. 13-16.

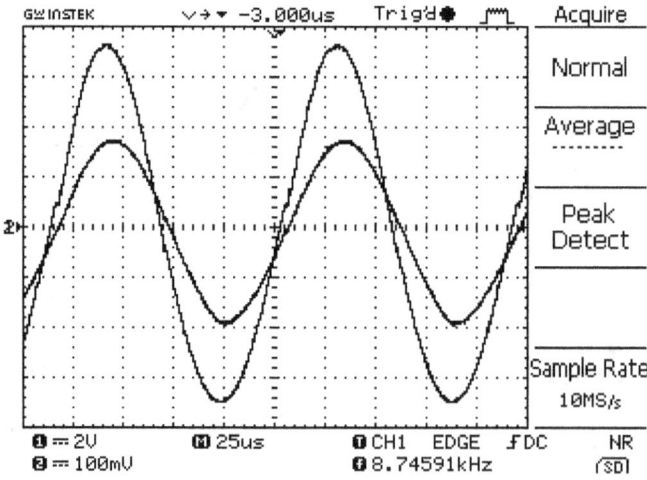

Fig. 13-16: TX and RX Signals (Garrett ACE Coil) with Ferrite Target

From this simple experiment you can now clearly see how it becomes possible to distinguish between non-ferrous versus ferrous, and also any response due to the ground.

Step 12: Sample Pulse Generators

Fit R21, R22, R26, R27 (10k), RV1, RV2, RV4, RV5 (10k preset), C13, C15 (100nF), C12, C14, C16, C17 (10nF), TP6, TP7, TP8, TP10 (test point), Q2, Q3 (2N3906), and U5, U6 (CD14583BE).

Temporarily insert jumpers across J2 (Ground Balance) and J4 (DISC) connectors. These external pot connections will be made later on in the build process.

Connect channel 1 scope probe to TP7 (GB Sample) and adjust RV2 to give a pulse width of 57us. This is equal to one half-period of the TX frequency of 8.75kHz.

Repeat this procedure for TP10 (DISC Sample) and adjust RV5.

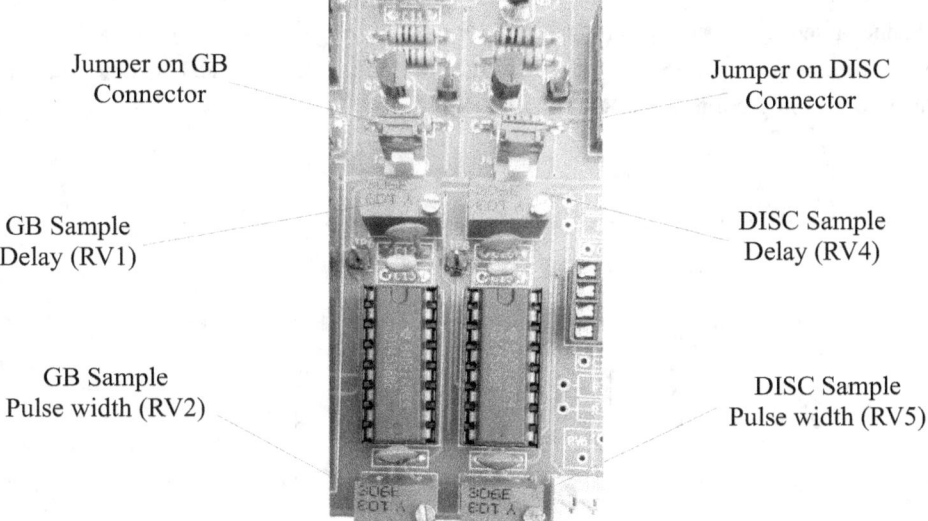

Fig. 13-17: Sample Pulse Generators

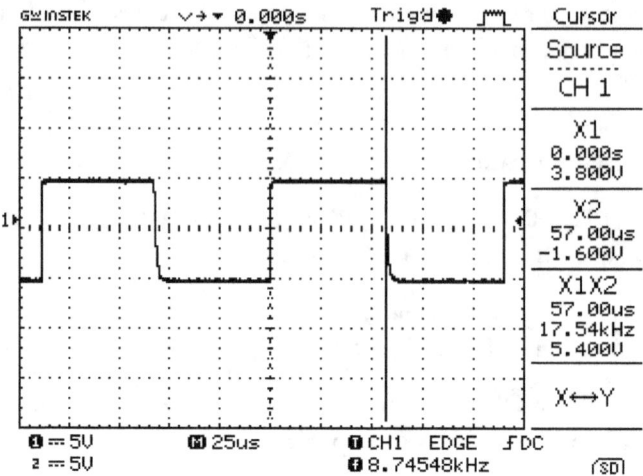

Fig. 13-18: Sample Pulse Width (57µs)

Next attach channel 1 scope probe to TP11 (preamp output) and trigger the scope on this signal. Connect channel 2 scope probe to TP7 (GB Sample), and adjust RV1 to position the GB Sample pulse over the negative-going zero-crossing point of the RX signal. See Fig. 13-19.

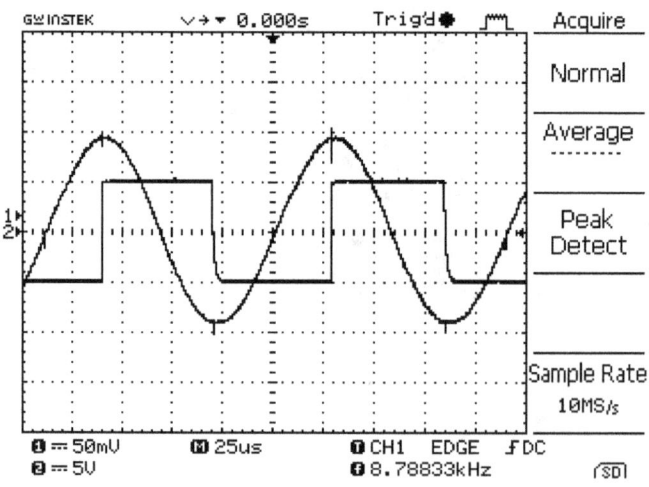

Fig. 13-19: GB Sample Pulse (Garrett ACE coil)

Repeat this procedure with channel 2 scope probe connected to TP10 (DISC Sample), but this time adjust RV4 to position the DISC Sample pulse over the negative part of the RX signal. See Fig. 13-20.

Step 13:   Synchronous Demodulators

Fit R37, R38 (20k), R49, R65 (100k), C19, C20, C25, C26 (100nF), C29, C40 (10nF), TP12, TP13 (test point), and U7 (4066).

Examine the signal at TP13 while moving a non-ferrous target towards the coil. The DC level of the signal should increase. Next do the same thing with a ferrous target and note that the signal also increases. Lastly move a ferrite core towards the coil. In this case you will most likely see either an increase or decrease because the approximate adjustment of RV1 that you made in Step 12 (Fig. 13-19) needs some refinement. Carefully adjust RV1 until the ferrite core is ignored. The detector will then be ground balanced.

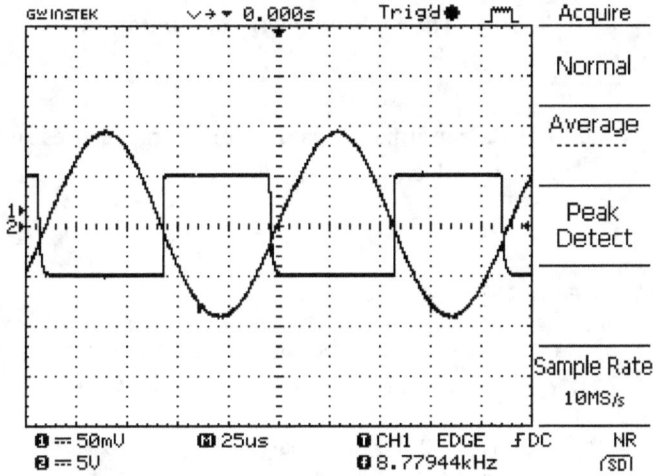

Fig. 13-20: DISC Sample Pulse (Garrett ACE Coil)

Fig. 13-21: Synchronous Demodulators

With the scope probe attached to TP12, confirm that a non-ferrous target causes an increase, whereas both a ferrous target and a ferrite core cause a decrease.

Step14:   First Differentiator Stage

Fit R48, R52, R63, R67 (470k), R50, R64 (47k), R51, R66 (56k), C27, C38 (470nF), C28, C39 (2n2), C34, C35 (100nF), TP15, TP16 (test point), and U10 (TL072).

Connect channel 1 scope probe to TP15 (GB) and channel 2 to TP16 (DISC). These test points are connected to the outputs of the first differentiator stages.

Then slowly pass a non-ferrous target over the coil and note that both the GB and DISC signals increase, then decrease before returning to their original DC levels. This happens because the differentiation circuits are tracking the rate-of-change of the signal. A ferrous target produces the same result in the GB signal, whereas the DISC signal decreases followed by an increase. A ferrite core should produce no change for GB, and give the same response as a ferrous target for DISC. It is quite likely you will have to slightly tweak the GB delay setting (RV1) to reject the ferrite.

Figures 13-22 to 13-24 show the results of swinging the non-ferrous, ferrous, and ferrite targets across the face of the coil. The scope was set to a slow timebase of 500ms in order to capture these events.

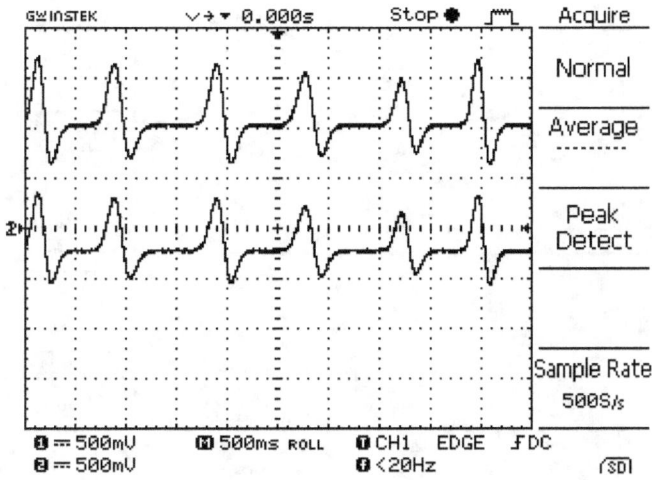

Fig. 13-22: Non-ferrous Target - 1ˢᵗ Diff (DISC at Top, GB at Bottom)

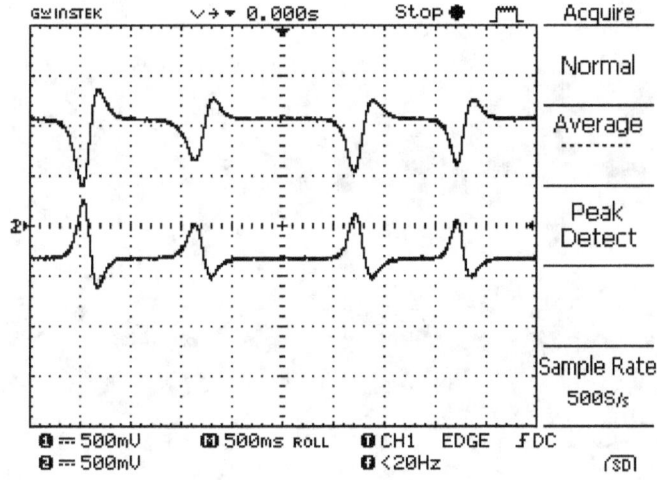

Fig. 13-23: Ferrous Target - 1ˢᵗ Diff (DISC at Top, GB at Bottom)

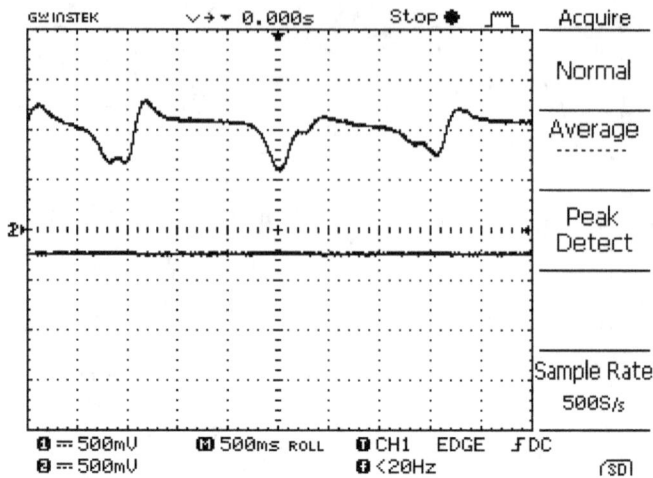

Fig. 13-24: Ferrite – 1st Diff (DISC at Top, GB at Bottom)

Fig. 13-25: First Differentiator Stage (GB and DISC)

Step 15: Second differentiator stage

Fit R53, R68 (22k), R56, R71 (100k), R55, R70 (10k), R54, R69 (1M), C36, C37 (100nF), C30, C41 (1µF), C31, C42 (10nF), TP14, TP17 (test point), and U11 (TL072).

Connect channel 1 scope probe to TP14 (GB) and channel 2 to TP17 (DISC). These test points are connected to the outputs of the second differentiator stages.

Then slowly pass a non-ferrous target over the coil and note that both the GB and DISC signals initially decrease, then increase followed by a second decrease before returning to their original DC levels. Due to the second differentiation circuit in each channel, the outputs are now tracking the rate-of-change of the rate-of-change of the target signal. A ferrous target produces the same result in the GB signal, whereas the DISC signal is inverted. It is not easy to reject the ferrite target using visual means (as in step 14) because of the extra amplification and increased sensitivity, so we will leave the final ground balancing adjustment until the audio stage is installed.

Figures 13-26 and 13-27 show the results of swinging the non-ferrous and ferrous targets across the face of the coil. As before, the scope was set to a slow timebase of 500ms in order to capture these events.

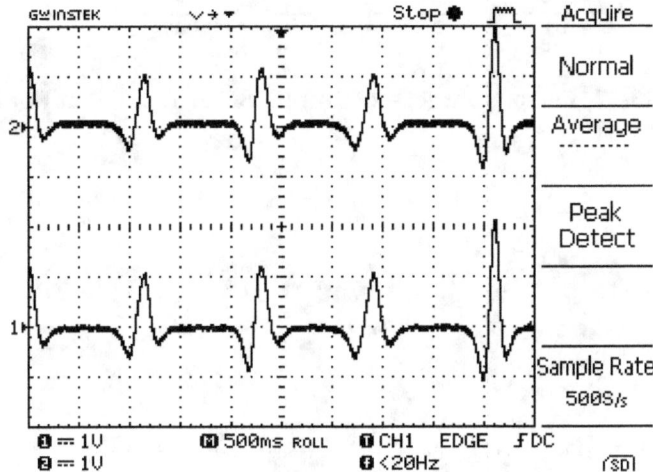

Fig. 13-26: Non-ferrous Target - 2nd Diff (DISC at Top, GB at Bottom)

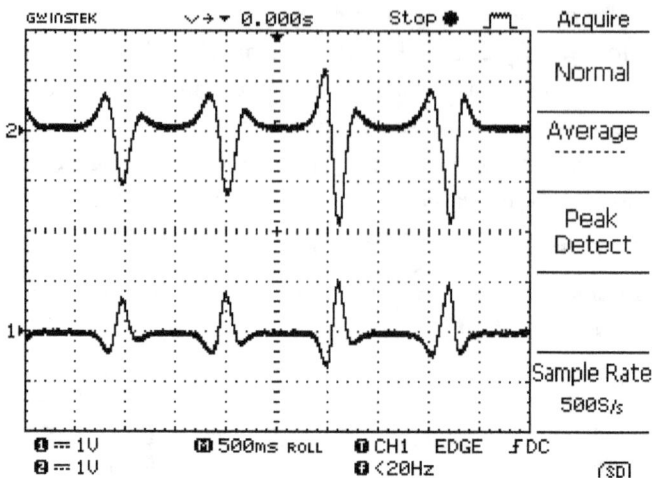

Fig. 13-27: Ferrous Target - 2nd Diff (DISC at Top, GB at Bottom)

Fig. 13-28: Second Differentiator Stage (GB and DISC)

Step 16: Comparators

Fit R33, R34 (4k7), R32, R35 (1M), R39 (3k3), RV6 (10k preset), C21, C22 (100nF), and U8 (LM393).

Wire an external 1k potentiometer for the Sensitivity control to connector J5. See Fig. 13-29 for wiring details. Note there is an additional 100R resistor in series with the SENS pot which is not included on the PCB.

Fit a jumper across J7 in place of the AM/DISC switch.

Attach channel 1 scope probe to the junction of R39 and D3 cathode (sorry, no test point provided). Set the Sensitivity control to maximum, and adjust RV6 (Sensitivity preset) to mid position. A non-ferrous target will cause the signal to jump to +5V, whereas there should be no response from a ferrous target. Removing the jumper from J7 will enable the All Metal (AM) mode of operation.

It is difficult to adjust the Sensitivity preset correctly at this stage, so final adjustment will be performed after installing the audio amplifier circuitry.

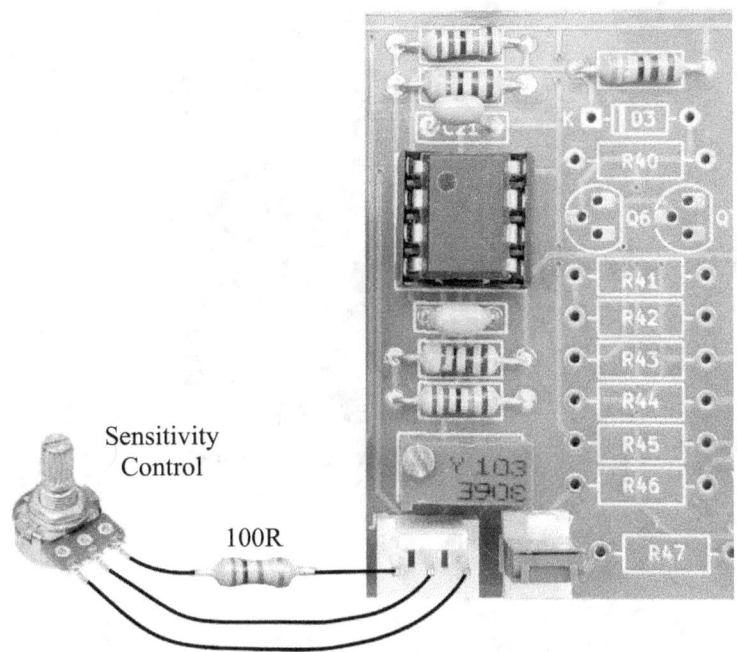

Sensitivity Control

100R

Fig. 13-29: Comparators and Sensitivity Control Connections

Step 17:   Audio Amplifier

Fit R40 (22k), R45, R46, R47 (10k), R41, R42 (100R), R43 (4k7), R44 (20R), C33 (1000μF), D3 (1N4148), Q6, Q8, Q9 (2N3904), and Q7 (2N3906).

Connect either a loudspeaker or headphones to J9. The detector should now beep when a non-ferrous target is passed in front of the coil, even though we have not yet calibrated the sensitivity control.

Sensitivity Control Calibration

Set the SENS control to maximum. Attach a scope probe to the SENS control pin that is connected to the 100R resistor, and adjust RV6 (SENS preset) so that the DC voltage is approximately 50mV. At this point the detector will have a lot of chatter on the audio output. Adjust RV6 to increase the voltage until the audio is on the limit of chattering. The DC voltage will probably be somewhere between 50mV and 60mV. That's all there is to the calibration procedure, and you can now carry out some air tests at different settings of the SENS control.

Discrimination may not be perfect at this point as the sample pulse was only positioned visually in step 13 (Fig. 13-19). This will be corrected in the next step.

Fig. 13-30: Audio Amplifier

Step 18:   External GB and DISC Controls

Both the external GB and DISC control connectors, J2 and J4 respectively, are two-pin. These pins need to be connected to the GB and DISC potentiometers so that the resistance increases as they are turned clockwise.

First remove the jumper from J2 and connect the GB control, setting it to mid-position. While listening to the audio output, adjust RV1 to reject a ferrite core. The GB control can now be used to eliminate ground response in the field.

Next remove the jumper from J4 and connect the DISC control, setting to minimum. That is, fully anti-clockwise. Then adjust RV4 to reject a ferrous (iron) target, but still accept foil. The DISC control can now be adjusted from the minimum position to provide discrimination of higher conductivity items such as foil, bottle caps, or pulltabs.

However (if you wish) you can leave step 18 until after the Arduino Nano #2 has been installed and a decision made on which type of display to use. This will also make it easier to test the detector on the bench without the external GB and DISC controls connected.

Step 19:   Arduino Nano #2

Fit R58, R59, R60, R61, R62 (10k), R57 (20k), C32 (10µF), TP18 (test point), and A2 (Arduino Nano).

Note correct orientation for the Arduino Nano with the USB connector at the lower edge of the PCB.

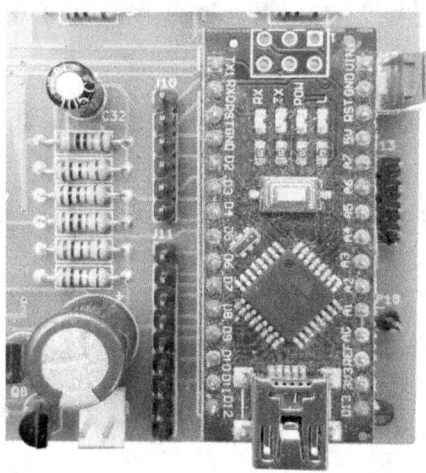

Fig. 13-31: Arduino Nano #2

Install a jumper across J12 (AM/DISC switch B). This is the second half of the double-pole double-throw (DPDT) all-metal/discrimination switch, and is used to tell the second Nano which particular mode is active. This information is then used to change the display information accordingly.

We will now examine in detail how to connect each of the displays mentioned in Chapter 11.

Please refer to Chapter 12 for details of the software for each display.

2-line 16-character LCD

This display should be wired to connector J11. The pin numbering starts with pin 1 at the top of the connector (see Fig. 13-31).

The pin mapping between J11, the LCD, and Arduino Nano is shown in Table. 13-1.

Fig. 13-32: 2-line 16-character LCD

LCD pins 7, 8, 9, and 10 (DATA0 to DATA3) are not used because the display is being driven in 4-bit mode.

If the LCD has a backlight, then an additional switch may be fitted to connect 0V to LCD pin 16 (BL-), and +5V to pin 15 (BL+) via a 100R resistor. To enable the contrast to be adjusted, a 10k preset is also required. This preset connects between +5V and 0V, with the middle pin going to LCD pin3.

J11 pin number	LCD pin number	LCD pin name	Arduino Nano pin number	Arduino Nano pin name
1	14	DATA7	9	D6
2	13	DATA6	10	D7
3	12	DATA5	11	D8
4	11	DATA4	12	D9
5	6	EN	13	D10
6	5	RW	14	D11
7	4	RS	15	D12
8	2	VDD		+5V
9	1	GND		0V

Table. 13-1: Pin Mapping Between J11, LCD, and the Arduino Nano

4-line 16-character LCD

The 4-line LCD uses the exactly the same connections as the 2-line version. Please refer to Table. 13-1.

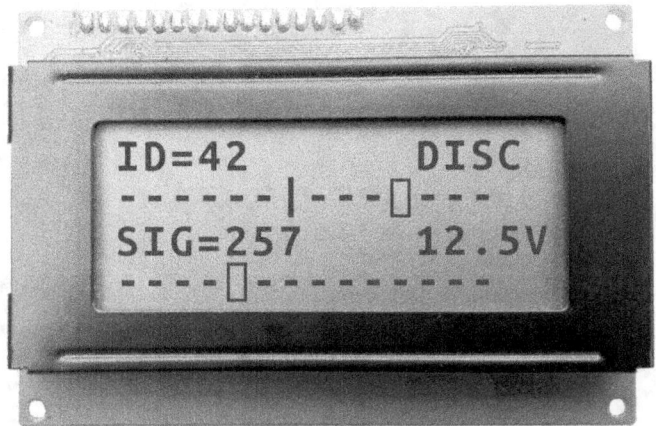

Fig. 13-33: 4-line 16-character LCD

SH1106 OLED 1.3" (128x64) Display

This display should be wired to connector J13. The pin numbering starts with pin 1 at the top of the connector (see Fig. 13-31).

The pin mapping between J13, the OLED, and Arduino Nano is shown in Table. 13-2.

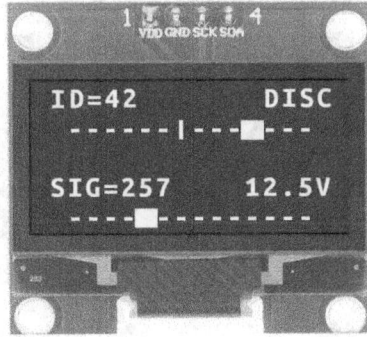

Fig. 13-34: SH1106 OLED 1.3" (128x64) Display

J13 pin number	OLED pin name	Arduino Nano pin number	Arduino Nano pin name
1	VDD		+5V
2	GND		GND
3	SCL	24	A5
4	SDA	23	A4

Table. 13-2: Pin Mapping Between J13, OLED, and the Arduino Nano

The OLED is the simplest display to connect hardware-wise, but communication is slower than with the LCDs due to the serial nature of the interface.

TFT 2.4" (320x240) Display

This display should be wired to connector J11. The pin numbering starts with pin 1 at the top of the connector (see Fig. 13-31).

The pin mapping between J11, the TFT, and Arduino Nano is shown in Table. 13-3.

Fig. 13.35: TFT 2.4" (320x240) Display

J11 pin number	TFT pin name	Arduino Nano pin number	Arduino Nano pin name
1	LED	9	D6
2	RESET	10	D7
3	DC	11	D8
4	CS	12	D9
5	SCK	13	D10
6	MOSI	14	D11
7	MISO (not used)	15	D12
8	VDD		+5V
9	GND		0V

Table. 13-3: Pin Mapping Between J11, TFT, and the Arduino Nano

Figure 13-35 shows the TFT in operation. Since the diagrams in this book are monochrome, the image is unable to show that the TFT is a colour display. However, the rear of the book cover shows how the display looks in full colour. For reference, the target ID will be either green or red depending on whether the number is positive or negative. The VDI bar at the top of the screen displays a cursor that is green if on the right-side, or red if on the left. The detector mode is displayed above the battery voltage, and will be either DISC or AM. This text is in yellow. The battery voltage is green if above or equal to 10.5V, and red if below. The operation of the DEPTH and CONF bars are explained in detail in Chapter 12. The DEPTH bar is always white, but the CONF bar is green if it's in the top half, and red in the bottom half.

Step 20:   Battery Supply

Up to this point you will have either been using a bench power supply or a battery pack during testing, but we now need to decide on a suitable portable rechargeable battery pack to incorporate into the detector.

Since this design has two Arduino Nanos, and the transmitter uses a forced oscillator, you may expect the current consumption to be relatively high. In fact the current consumption is fairly acceptable at 240 to 250mA if used with a Li-ion battery pack. For example, the pack we will be using here (see Fig. 13-36) has a nominal voltage of 11.1V and a capacity of 4800mAh. An average current consumption of 250mA will therefore provide an estimated detecting time of 19.2 hrs.

Fig. 13-36: Rechargeable Li-ion Battery Pack

One quirk of this particular power pack is that the on/off switch on the end of the enclosure needs to be in the ON position while charging, and since there is an LED built into the switch we cannot simply leave the switch in the ON position as the LED will drain the battery even when the detector is turned off. This means that we need to carry out a small modification. Obviously this is at your own risk, as it does say on the label: "Do not attempt to open disassemble or service the battery pack".

The modification is actually quite simple. If you remove the four screws that secure the case it is highly likely that the case will not open fully, as the batteries appear to be secured with some kind of adhesive. However, it is only necessary to prise apart the end where the switch is located. This creates enough room to extract the switch so that you can examine the wiring. With reference to Fig. 13-37, there are three pins on the switch, and one of them is used to provide a connection to the built-in LED. This pin has two black wires soldered to it. The other two pins have one red wire each. When the switch is in the ON position, the red wires are connected together. Although the switch looks like an SPDT, it is in fact an SPST. The pin with the two black wires is never connected to any of the other pins, and only serves to light the LED.

The modification requires the removal of the two black wires. These wires must be soldered together and covered in heatshrink sleeving. This effectively disconnects the LED and prevents the batteries from discharging when the detector is turned off. The battery can then be reassembled and the switch left in the ON position.

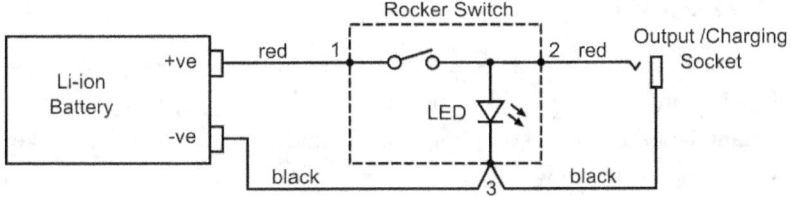

Fig. 13-37: Li-ion Battery Pack Wiring Before Modification

Acknowledging the old adage that a picture is worth a thousand words, please compare figures 13-37 and 13-38. This should make it abundantly clear why this modification is required.

Lastly, we need to connect the battery pack to the detector, on-off switch and charging connector. The suggested wiring is shown in Fig. 13-39.

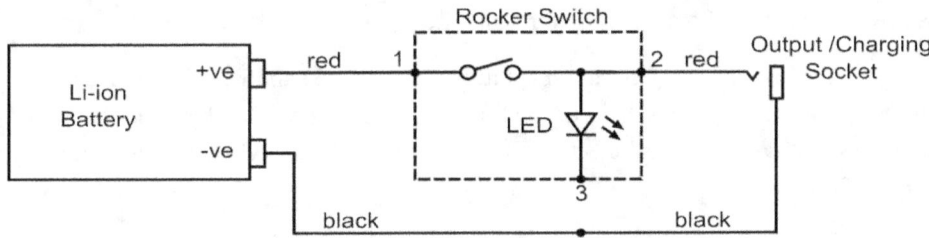

Fig. 13-38: Li-ion Battery Pack Wiring After Modification

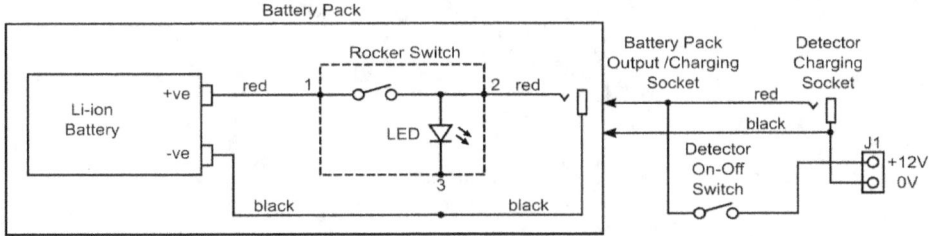

Fig. 13-39: Detector Power Supply Wiring

Step 21:   Coil Socket Wiring

Although the Arduino VLF detector can use a multitude of coils from several detector manufacturers, the coil connections they use are invariably different. However, in practice, it will be necessary to make a decision on which manufacturer's coils you prefer to use, and then wire the coil socket accordingly, or alternatively use your own proprietary wiring scheme.

For this particular build example I used the coil socket connections shown in Fig. 13-40, together with a Garrett ACE 9" x 6.5" elliptical concentric coil. This meant that I had to change the coil plug wiring to match the chosen socket. You may not want to do this yourself as it means the coil can no longer be used with a Garrett ACE detector. However, as I wanted to carry out a series of experiments, this made it easier to swap coils.

The Garrett ACE 200 has a TX frequency of 6.5kHz and a default coil size of 9" x 6.5". The ACE 300's TX frequency is 8kHz with a default coil size of 10" x 7", and the ACE 400's TX frequency is 10kHz with a default coil size of 11" x 8.5". Although these are the standard coils for each model, all the Garrett ACE coils are compatible with all models. Therefore you will be able to use any of these coils with the Arduino Nano VLF detector without having to recalibrate.

With reference to Fig. 13-11, the correct pin mapping to the TX (J3) and RX (J6) connectors on the PCB for our proprietary coil socket wiring has changed from 1 4 2 3 to 4 5 3 2. The screen wire in the Garrett ACE coil plug was wired to pin 1 and electrically connected to the TX GND at pin 4 on the back of the socket. (see Fig. 13-40).

This proprietary coil socket wiring allows for easier experimentation with different coils, and also handles situations where the cable screen is independent from the TX and RX ground wires.

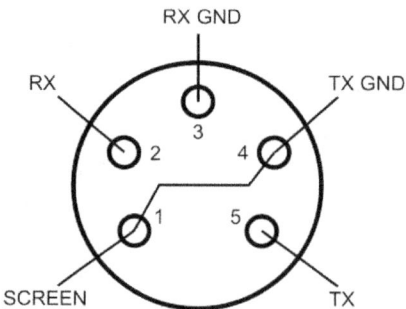

Fig. 13-40: Rear View of Coil Socket (Proprietary)

Step 22: All-Metal / DISC Switch

The AM/DISC mode switch wiring is shown in Fig. 13-41. The prototype unit used a DPDT switch, but an SPDT would also have been suitable. When the switch is in the down position this enables DISC mode, which connects the pins of J12 together to allow the DISC channel to disable the audio when a ferrous target is discriminated. The connector J7 is directly connected to a digital input on the second Arduino Nano that is pulled high by a software-enabled pull-up resistor. When the pins of J7 are shorted together, this pulls the input low and instructs the processor that the detector is in DISC mode.

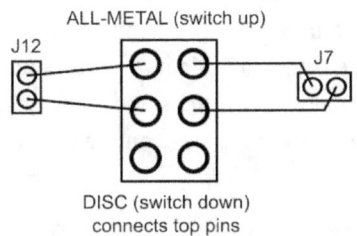

Fig. 13-41: AM/DISC Switch Wiring

Step 23: Mechanical Construction

If you've managed to reach this point, then it's time to put the electronics in a suitable enclosure and test it outside in the real world.

Luckily there are many detector parts available nowadays that can be purchased for a modest price, or you may be able to scavenge parts from a broken detector bought on an auction site.

For the prototype built by the author, a strong fibreglass straight stem from a Nexus detector was used, together with an arm cup from White's Electronics (who sadly ceased trading in June 2020 after being in business since 1950). The control box (located under the arm cup) was from RS Components (Stock number: 1920756) made from ABS plastic with front and rear panels providing IP53 protection against liquids and solid objects. The enclosure size was height: 60mm, width: 136mm, and length: 185.5mm.

The control box was used to house the PCB and lithium power pack. The user controls (GB, DISC, Threshold, and AM/DISC switch) were positioned on the rear panel. The ON/OFF switch, and the coil, headphone, and charging sockets were placed on the front panel. The controls were placed on the rear panel (instead of on the top) to prevent them from being accidentally adjusted while using the detector, and to make it easier to operate and calibrate the detector on the bench when disassembled.

A second smaller enclosure was mounted on top of the handle to house the TFT display. This was a Hammond enclosure (part number: 1553) also purchased from RS Components.

There were a few 3D-printed parts used to mount the control box on the stem, and to mount the display enclosure on top of the handle.

Of course, how you decide to build your own detector is entirely up to you.

Step 24: Testing the Detector

Here we come to the inevitable question that almost everyone asks: "How deep does it go?".

This is a difficult question to answer because there are many different factors that make direct comparisons somewhat tricky. For example, ground conditions, coil sizes, detector type, etc., etc. I'm sure you've seen videos online for detectors that demonstrate fantastic depths on a test target buried in areas with no trash. However, if you were to test the same detector in an area infested with ferrous trash the results would be very disappointing. Many of these online tests are also performed on the bench using what is known as an *air-test*.

Continuous wave detectors (commonly referred to as VLF, or *very low frequency*) will generally perform much better in an air-test than they do on the same targets buried in the ground, whereas pulse induction detectors demonstrate little or no difference between air and ground tests.

In my test garden there are 5 English decimal pennies buried in a line and placed 1 metre apart. The pennies are all the same date, and without an iron core. The first is buried at 2", and the next at 3", etc., up to 6". Each penny was epoxied to the end of a length of wood dowelling and hammered into the ground. Although there is nothing but the dowelling directly above each coin, the garden is literally infested with pieces of ferrous trash such as small nails. This makes the detection of the 5 pennies very challenging. Using a PI detector in the test garden is like listening to a machine gun (unless you're using the Voodoo PI detector - see Appendix E).

Virtually all detectors can detect the first coin, except for one particular inexpensive TR detector that is still being sold today by the manufacturer. The second coin is also detectable by the majority of detectors with some exceptions. To be honest the signal from the third coin is very *iffy* and no-one would decide that it was a good target. The 4th and 5th coins are never detected by any detector. The only occasion when I might have thought the 3rd coin was worth digging, was while using a 4" diameter Tesoro coil with the Raptor detector from ITMD. In Chapter 14 the 4" Tesoro coil is one of test subjects, so you can see how it performs with the Arduino Nano VLF detector.

Bearing in mind all the caveats above, the following air-test results were obtained with the Garrett ACE 9" x 6.5" coil:

Victorian Penny:	9"
1 Euro:	7"
U.S. Nickel:	7"
U.K. Decimal Penny:	6"
Hammered Silver:	6"
Cola Can:	13"

"One good test is worth a thousand expert opinions."

--- Wernher von Braun

In Chapter 13 a Garrett ACE coil was used for the initial testing, and in this chapter we will be exploring the use of various coils from different metal detector manufacturers. It was stated previously that virtually any induction balanced search head can be used with this design, with the proviso that it contains one TX coil and one RX coil, without any additional components that would prevent the TX coil being force driven or the RX coil from being tuned with an external capacitor. Theoretically this is correct, but as often happens in practice the reality may be somewhat different. So, with no further ado, let's start the testing and find out ...

General Calibration Procedure

In order to calibrate the detector for the various coils tested, the following general procedure was followed:

1. Measure the TX and RX inductances and calculate a suitable value for the RX tuning capacitor (as described in Chapter 13 Step-10) based on the desired TX frequency.

2. Re-wire the coil plug, and connect the coil. Refer to Fig. 13-40 and Fig. 14-2.

3. Using an oscilloscope, note how the amplitude and phase of the RX waveform at the preamp output (TP11) changes relative to the TX signal (TP9).

4. Switch the detector to AM (all-metal) mode and set the external GB control to mid-position.

5. Monitor the RX signal at the preamp output (TP11) versus the GB sample pulse (TP7), and adjust the GB sample pulse width trimmer (RV2) to half the period of the TX frequency.

6. Adjust the GB sample pulse delay trimmer (RV1) to the appropriate zero-crossing position on the RX waveform. At this point the detector should beep for both non-ferrous (for example, a coin) and ferrous (iron).

7. Make fine adjustments to the GB sample delay to eliminate a ferrite slug while still reacting to non-ferrous and ferrous targets. The detector is now ground balanced.

8. Switch the detector to DISC (discrimination) mode and set the external DISC fully anti-clockwise.

9. Monitor the RX signal at the preamp output (TP11) versus the DISC sample pulse (TP10), and adjust the DISC sample pulse width trimmer (RV5) to half the period of the TX frequency.

10. Adjust the DISC sample pulse delay trimmer (RV4) to the appropriate peak on the RX waveform. At this point the detector should beep for non-ferrous targets and reject ferrous targets. If ferrous targets are accepted and non-ferrous are rejected, then the DISC sample pulse is on the wrong peak.

Note that you may have to modify the sketch for the Arduino Nano #1 to allow the GB and DISC sample pulses to be positioned on the correct zero-crossings and peaks of the RX waveform. For the coils examined in this chapter the Arduino sketches are available on the Geotech website as well as being listed here.

Tesoro 4" Concentric (5-pin)

Some of my favourite coils to use for home-brewed metal detector designs are the ones manufactured by Tesoro. Unfortunately the company went out of business in 2018, but there are still many coils available second-hand. These coils are relatively simple to get working, have a lightweight construction, provide excellent iron rejection, and are easily ground balanced.

A typical transmit frequency for Tesoro detectors is 10kHz or thereabouts. The actual frequency is not that critical.

There were a few changes required to the software for the Arduino Nano #1 to change the TX frequency to approximately 10kHz. The DISC sample point was also moved to align it with the positive peak of the RX waveform. Note the addition of a couple of 62.5ns delays implemented by calling an assembler NOP command.

Fig. 14-1: Tesoro 4" Concentric Coil

```
// Arduino Nano VLF metal Detector
// Nano #1 with sinewave output
// 10kHz for Tesoro coils

// Pin assignments
byte audioPin = 11;         // Audio tone

// Program variables
float deg, rad;             // Degrees and radians
float tx;                   // TX value for each data point
byte dacInput[180];         // R-2R
byte syncState = LOW;       // State of SYNC pulse (HIGH or LOW)

// Power supply sync pulse
void sync() {
  if (syncState == LOW) {
    PORTB |= B00000100;     // If SYNC state is LOW, then toggle B2 HIGH
    syncState = HIGH;       // Save current SYNC state
  } else {
    PORTB &= B11111011;     // If SYNC state is HIGH, then toggle B2 LOW
    syncState = LOW;        // Save current SYNC state
  }
}

// GB sample trigger
void gbTrigger() {          // Output GB trigger pulse
  PORTB &= B11111110;       // Set B0 LOW
  PORTB |= B00000001;       // Set B0 HIGH
}

// DISC sample trigger
void discTrigger() {        // Output DISC trigger pulse
  PORTB &= B11111101;       // Set B1 LOW
  PORTB |= B00000010;       // Set B1 HIGH
}

void setup() {
  // Calculate DAC input values
  deg = 0;                              // Start with zero degrees
  for (int i = 0; i <= 179; i++) {      // Sine wave will consist of 180 steps
    rad = deg / 360 * TWO_PI;           // Convert degrees to radians
    tx = sin(rad) * 2.5 + 2.5;          // Scale and offset
    // 5V divide by 255 = 0.0196
    // R/2R ladder DAC resolution is 8-bits (19.6mV)
    dacInput[i] = tx / 0.0196;          // Save value for current step
    deg += 2;                           // Increment by 2 degrees
  }
  for (int i = 0; i <= 11; i++) {       // Set D0 to D11 as outputs
    pinMode(i, OUTPUT);
  }
  for (int i = 0; i <= 10000; i++) {    // Drive SYNC pulse for 1 second
    sync();
    delayMicroseconds(100);
  }
  noInterrupts();                       // Disable interrupts
  analogWrite(audioPin, 127);           // Set audioPin with 50% duty cycle PWM
}

void loop() {
  sync();                               // Sync voltage converters
  for (int i = 0; i <= 89; i++) {       // Apply positive half of sine wave to DAC inputs
    PORTD = dacInput[i];
    asm("nop\n");                       // Add a 62.5ns delay
  }
  gbTrigger();                          // GB sample trigger
  discTrigger();                        // DISC sample trigger
```

```
    for (int i = 90; i <= 179; i++) {     // Apply negative half of sine wave to DAC inputs
        PORTD = dacInput[i];
        asm("nop\n");                      // Add a 62.5ns delay
    }
}
```

There are two types of coils that Tesoro have supplied over the years, which are easily identified by the number of pins in the connector. The 4" coil tested here is of the 5-pin variety, and has a transmit loop inductance of 6mH and a receive inductance of 6.5mH. We'll get to the 4-pin types later on in this chapter.

Only one simple change to the hardware was required (i.e. fitting a 33nF capacitor to the RX tuning capacitor socket J8) and changing the GB and DISC sample pulse widths to 50µs by adjusting the RV2 and RV5 trimmers. The coil plug was re-wired (see Fig. 14-2) to match our proprietary setup shown in Fig. 13-40. Note that 5-pin Tesoro coils have a connection in the coil shell between RX GND and the cable screen (shown by a dotted line).

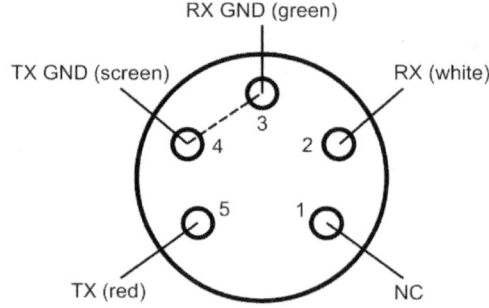

Fig 14-2: Tesoro 4" Concentric Coil Connections (Rear View of Plug)

GB sampling at the positive-going zero-crossing produced a reaction to both non-ferrous and ferrous targets while ignoring ground response (ferrite). Iron discrimination was then provided by sampling at the positive peak where only non-ferrous targets were detected.

Considering that the coil is only 4" in diameter, the following air-test results are extremely good:

```
        Victorian Penny:          7"
        1 Euro:                   6"
        U.S. Nickel:              7"
        U.K. Decimal Penny:       5"
        Hammered Silver:          5"
        Cola Can:                11"
```

Tesoro 9" x 8" Elliptical Concentric (5-pin)

Another good thing about Tesoro coils is their consistency. It was therefore a simple matter of switching to the elliptical coil without having to recalibrate the detector or change the tuning capacitor. The coil plug wiring required was also identical to the 4" coil shown in Fig. 14-2.

The following air-test results were obtained:

```
Victorian Penny:        11"
1 Euro:                 10"
U.S. Nickel:             8"
U.K. Decimal Penny:      8"
Hammered Silver:         8"
Cola Can:               18"
```

Fig. 14-3: Tesoro 9" x 8" Elliptical Concentric (5-pin) Coil

Tesoro 12" x 10" Elliptical Concentric (5-pin)

This is the largest of the Tesoro 5-pin coils tested in this chapter. Again it was a simple matter of connecting the coil without having to recalibrate the detector or change the tuning capacitor. In this case the wiring was identical to Fig. 14-2, and only the colours were different.

Fig. 14-4: Tesoro 12" x 10" Elliptical Concentric (5-pin) Coil

The following air-test results were obtained:

```
Victorian Penny:        12"
1 Euro:                 11"
U.S. Nickel:            10"
U.K. Decimal Penny:      9"
Hammered Silver:         8"
Cola Can:               20"
```

Note that there was some improvement in the air-test results compared to the 9" x 8" coil. The larger diameter also allows for increased ground coverage while searching, but may reduce sensitivity to smaller targets.

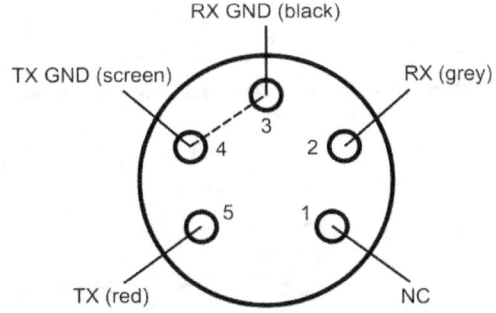

Fig 14-5: Tesoro 12" x 10" Concentric Coil Connections (Rear View of Plug)

Tesoro 5 ¾" Widescan DD (4-pin)

Fig. 14-6: Tesoro 5.75" Double-D Widescan (4-pin) Coil

This widescan coil from Tesoro is designed to work better in bad soil and to handle ground mineralization better than a concentric coil.

Tesoro manufactured this coil with a 4-pin plug to distinguish it from the 5-pin coils. This is because the TX and RX inductances are completely different. The TX inductance was measured as 930µH and the RX inductance as 15.8mH.

The Arduino Nano #1 software did not require any modification, and only the RX tuning capacitor needed changing to 15nF. The coil plug wiring is shown in Fig. 14-6. Note that the TX and RX screens are not connected in the coil shell, so it was necessary to add a wire link inside the plug to connect pins 3 and 4 together.

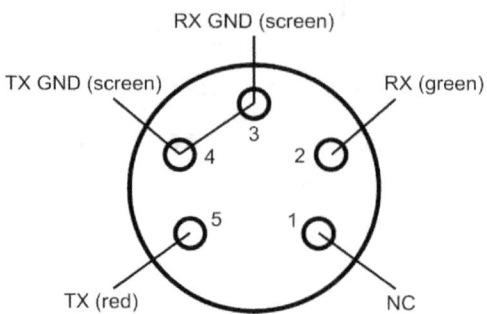

Fig 14-7: Tesoro 5.75" Widescan Coil Connections (Rear View of Plug)

The following air-test results were obtained:

```
Victorian Penny:       11"
1 Euro:                11"
U.S. Nickel:            9"
U.K. Decimal Penny:     9"
Hammered Silver:        8"
Cola Can:              18"
```

Tesoro 9" x 8" Elliptical Concentric (4-pin)

Fig. 14-8: Tesoro 9" x 8" Elliptical Concentric (4-pin) Coil

This coil looks identical in construction to the Tesoro 9" x 8" elliptical concentric except for the 4-pin plug. However, like the widescan coil, the TX and RX inductances are completely different. The TX inductance was measured as 930µH and the RX inductance as 15.3mH.

The Arduino Nano #1 software did not require any modification, and only the RX tuning capacitor needed changing to 15nF. The coil plug wiring is shown in Fig. 14-9. Like the widescan coil, the TX and RX screens were not connected in the coil shell, and it was necessary to put a wire link between pins 3 and 4 of the coil plug.

The GB sample was positioned at the positive-going zero-crossing of the RX waveform, and the DISC sample at the positive peak

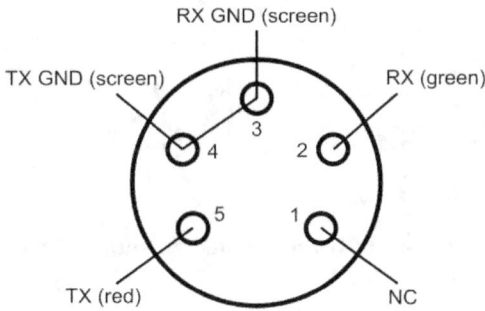

Fig 14-9: Tesoro 9" x 8" Elliptical Concentric Coil Connections (Rear View of Plug)

The following air-test results were obtained:

Victorian Penny:	12"
1 Euro:	11"
U.S. Nickel:	10"
U.K. Decimal Penny:	9"
Hammered Silver:	8"
Cola Can:	20"

and proved to be identical to those from the 9" x 8" (5-pin) coil.

Home-made 8" double-D

This coil is a home-made double-D coil whose construction is described in Chapter 10 of ITMD Edition 2.

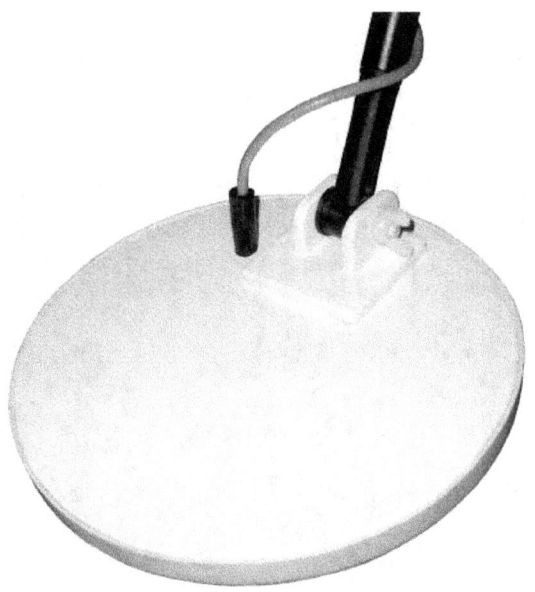

Fig. 14-10: Home-made 8" Double-D Coil

Since this coil was designed to mimic the characteristics of a 5-pin Tesoro coil, it was simply a matter of fitting a 33nF RX tuning capacitor and using the same settings as before.

The performance of this coil was reduced by an inch or two when compared to the Tesoro commercial coils. This was probably due to the TX and RX loops being shielded using aluminium tape. In retrospect it would have been better to have used adhesive copper tape.

Fisher F75 11" x 7" double-D

The TX frequency of the Fisher F75 detector is 13kHz, but in this test we will continue to use 10kHz for convenience. As you carry out your own testing of the coils in your collection it will soon become clear that there is often a lot of flexibility in the choice of TX frequency.

This particular coil is lightweight and is highly sensitive. It was also a very simple process to calibrate the detector and to obtain a stable ground balance setting with excellent iron rejection.

Fig. 14-11: Fisher F75 11" x 7" Double-D Coil

The TX inductance was measured as 670uH, and the RX inductance as 6.8mH. With a TX frequency of 10kHz, the RX tuning capacitor was kept as 33nF (the same as the 5-pin Tesoro coils). Obviously you could experiment further if desired to bring the RX coil closer to resonance and make further improvements in target depth, but I will leave that to the reader. The main purpose of this chapter is to explore how easy it is (or otherwise) to get different coils to work with the Arduino Nano VLF detector, not to push the detector to its limits.

The coil plug wiring is shown in Fig. 14-12.

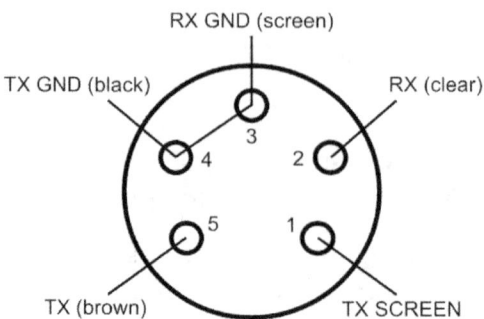

Fig 14-12: Fisher F75 11" x 7" Elliptical Double-D Coil Connections (Rear View of Plug)

Note that the TX screen is connected to pin1, which is then connected to pins 3 and 4 by a link on the detector's coil connector. Please refer to Fig. 13-40.

It was noted during testing that the the TX waveform was slightly distorted, and it was concluded that this was caused by the very low TX coil resistance of only 2.3 ohms. However, this did not appear to affect the performance of the detector. If it is decided to use a Fisher F75 coil permanently then it would be worth adding a 1 or 2 ohm resistor in series with the TX coil.

The GB sample pulse was positioned near the positive-going zero-crossing of the RX waveform, and the DISC sample pulse at the positive peak.

The following air-test results were obtained:

```
Victorian Penny:          14"
1 Euro:                   13"
U.S. Nickel:              12"
U.K. Decimal Penny:       10"
Hammered Silver:          10"
Cola Can:                 20"
```

A small change was also required to the Arduino Nano #1 sketch to trigger the DISC sample in the first part of the main loop.

```
void loop() {
  sync();                            // Sync voltage converters
  discTrigger();                     // DISC sample trigger
  for (int i = 0; i <= 89; i++) {    // Apply positive half of sine wave to DAC
inputs
    PORTD = dacInput[i];
    asm("nop\n");                    // Add a 62.5ns delay
  }
  gbTrigger();                       // GB sample trigger
  for (int i = 90; i <= 179; i++) {  // Apply negative half of sine wave to DAC
inputs
    PORTD = dacInput[i];
    asm("nop\n");                    // Add a 62.5ns delay
  }
}
```

Chinese (unbranded) 9.7" Concentric

I was very interested to test this particular coil with the Arduino Nano VLF design because these are available at an inexpensive price from China via a well-known online auction site. They are claimed to be compatible with the MD3030, MD4030 and MD4060 metal detectors. The coil was surprisingly light weight and very stable.

The TX inductance was measured as 1.3mH, and the RX inductance as 1.45mH. The cable screen is separate from the TX and RX wires, and there is a connection in the coil shell between the RX coil and the screen in the same way as the 5-pin Tesoro coils. The wire link on the back of the detector coil socket between pins 1 and 4 provides the required connection between one end of the TX coil and the 0V line.

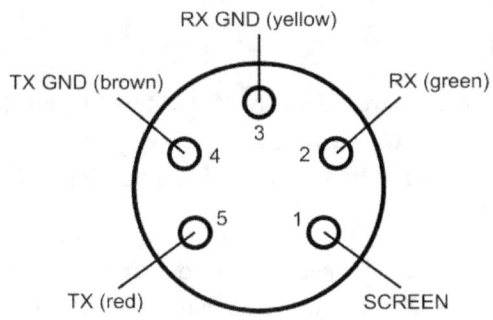

Fig 14-13: Chinese 9.7" Concentric Coil Connections (Rear View of Plug)

The same Arduino sketch that was used for the Fisher F75 was used in this test, with the TX frequency at 10kHz. The value for the RX tuning capacitor was calculated as 176nF, but the closest preferred value was 220nF.

Fig. 14-14: Chinese 9.7" Concentric Unbranded Coil

The GB sample pulse was positioned near the positive-going zero-crossing of the RX waveform, and the DISC sample pulse at the positive peak.

The following air-test results were obtained:

```
Victorian Penny:        11"
1 Euro:                 10"
U.S. Nickel:             7"
U.K. Decimal Penny:      7"
Hammered Silver:         7"
Cola Can:               17"
```

Conclusion

Although only a small number of coils were available for testing, the results were positive. In particular, the Fisher F75 coil gave the best air test results, and the unbranded Chinese coil was very stable and performed better than expected providing an inexpensive solution for home made designs. The Tesoro 4" concentric was the best for trashy sites.

Air tests can provide a relative indication of performance on various targets, but it must be noted that the actual numbers are somewhat subjective and you may obtain slightly different results.

Target	Garrett ACE 6.5" concentric	Tesoro 4" concentric 5-pin	Tesoro 9"x8" concentric 5-pin	Tesoro 12"x10" concentric 5-pin	Tesoro 5.75" widescan 4-pin	Tesoro 9"x8" concentric 4-pin	Fisher F75 11"x7" double-D	Chinese unbranded 9.7" concentric
Victorian Penny	9"	7"	11"	12"	11"	12"	14"	11"
One Euro	7"	6"	10"	11"	11"	11"	13"	10"
U.S. Nickel	7"	5"	8"	10"	9"	10"	12"	7"
U.K. decimal Penny	6"	5"	8"	9"	9"	9"	10"	7"
Hammered Silver	6"	5"	8"	8"	8"	8"	10"	7"
Cola Can	13"	11"	18"	20"	18"	20"	20"	17"

Table. 14-1: Results for All Coils Tested

Other coils

Unfortunately there were no Minelab coils readily available for testing, except for a Tornado DD which contains circuitry inside the coil shell and is therefore incompatible with this design.

The Garrett Crossfire coil was likewise of no use because it contains some passive components.

A Viking 6DX DD was also tested. It has a centre-tapped RX coil, which should be OK in theory, but it turns out that there are tuning capacitors already present in the coil shell.

I did have high hopes for a Golden Mask 7" DD coil, but after numerous attempts it was not possible to get this coil to behave. Non-ferrous targets caused a phase-shift left and a decrease in amplitude, whereas ferrous targets caused a phase-shift right and also a decrease in amplitude. As a result the GB sample could either be set to produce a beep on non-ferrous, and a double-beep on ferrous, or vice versa. This was a second-hand coil purchased from an online auction site, so it's possible it could have been faulty.

As the saying goes: YMMV (your mileage may vary).

Chapter 15

"In a dark place we find ourselves, and a little more knowledge lights our way."

--- Yoda

The goal of this project was to create a powerful and flexible design that could make use of a huge range of commercial metal detector coils. To achieve this flexibility it was decided at the start of the project to use the inexpensive and popular Arduino Nano processor. The idea was to make the project accessible to a wider group of hobbyists who perhaps were not familiar with the Microchip range of PIC processors (as used in the Voodoo Project).

Overall the design created in the previous chapters does meet the goals that were set out for the project. It is clear that several coils from different manufacturers can be used, along with several different types of display. In general, the design can easily be adapted by any competent hobbyist to create their own variants.

You may have been wondering why a decision was made to use a DAC to generate a sine wave for the TX signal. Would it not have been easier to simply output a square wave? Although the answer is "yes", transmitting a square wave generates an odd number of harmonics along with the fundamental. These harmonics waste transmit energy, and could potentially interfere with other nearby metal detectors. Also, generating a square wave output would just have been a less interesting approach. That's my excuse anyway.

Although using a second Arduino Nano allows the detector to use a multitude of displays, you may (for whatever reason) not wish to incorporate one when you build this design. In which case leaving out the second Nano and associated components will not affect its operation, but it just means that you will not have the capability of displaying possible target IDs or battery charge state. This is (of course) perfectly OK, and in fact many detector manufacturers have produced designs without a display of any kind.

"I guess the question I'm asked the most often is: "When you were sitting in that capsule listening to the count-down, how did you feel?" Well, the answer to that one is easy. I felt exactly how you would feel if you were getting ready to launch and knew you were sitting on top of two million parts — all built by the lowest bidder on a government contract."

-- John Glenn

ICs

Component	Names	Quantity	Pins
Arduino Nano V3.x	A1, A2	2	30
CD14538BE	U5, U6	2	16
4066	U7	1	14
LM393	U8	1	8
TL071	U9	1	8
TL072	U4, U10, U11	3	8

3-pin Voltage Regulators

Component	Names	Quantity
79L05	U1	1

Transistors

Component	Names	Quantity
2N3904	Q1, Q4, Q6, Q8, Q9	5
2N3906	Q2, Q3, Q5, Q7	4

Diodes

Component	Names	Quantity
1N4148	D1-D3	3

Resistors (1% tolerance)

Component	Names	Quantity	Spacing (thou)
10R	R23, R24	2	400
20R	R44	1	400
100R	R41, R42	2	400
1k	R36	1	400
3k3	R39	1	400
4k7	R33, R34, R43	3	400
5k1	R28, R30	2	400
10k	R3, R5, R7, R9, R11, R13, R15, R18, R21, R27, R19, R20, R22, R26, R45, R46, R47, R55, R58, R59, R60, R61, R62, R79	24	400
10k trimmer	RV1-RV6	6	N/A
18k	R25	1	400
20k	R1, R2, R4, R6, R8, R10, R12, R14, R16, R17, R37, R38, R57	13	400
22k	R40, R53, R68	3	400
47k	R50, R64	2	400
56k	R51, R66	2	400
100k	R29, R31, R49, R56, R65, R71	6	400
470k	R48, R52, R63, R67	4	400
1M	R32, R35, R54, R69	4	400

Capacitors

Component	Names	Quantity	Spacing (thou)
2n2	C28, C39	2	200
10nF	C12, C14, C16, C17, C29, C31, C40, C42	8	200
47nF	C27, C38	2	200
100nF	C9, C11, C13, C15, C19-C26, C34-37	16	200
220nF	C18	1	200
1µF	C10, C30, C41	3	200
10µF electrolytic	C2, C3, C5, C6, C32	5	100
470µF electrolytic	C1, C4, C7, C8	4	200
1000µF electrolytic	C33	1	200

Connectors and Headers

Component	Names	Quantity
2-pin connector	J1-J4, J6, J7, J9, J12	8
2-pin screw connector	J8	1
3-pin connector	J5	1
4-pin header	J11	1
6-pin header	J10	1
9-pin header	J13	1

Other

Component	Names	Quantity
Test point	TP1-TP18	18
GB external pot (1k)	Connects to J2	1
DISC external pot (1k)	Connects to J4	1
SENS external pot (1k)	Connects to J3	1
100R 1% resistor	In series with SENS pot (not on PCB)	1
AM/DISC switch	Connects to J7 and J9	1
2-pin plug	Required for J1-J4, J6, J7, J9	7
3-pin plug	Required for J5	1

Miscellaneous

Depending on the actual construction, there are several additional components required beyond the external potentiometers and those on the PCB itself.

For example:

- On-off switch

- Headphones (or high-impedance speaker)

- Headphone socket (recommended)

- Either LCD, OLED or TFT display

- IC sockets (optional)

- Electronic control box

- Battery pack (Li-ion pack, described in Chapter 13)

- Battery charger socket

- Coil connector (5-pin)

- Coil (either commercial or custom made)

- Coil cable (coaxial, RG58 or equivalent)

- Detector stem, arm cup, stem bolts, etc.

"Whenever you are asked if you can do the job, tell 'em, 'Certainly, I can.' Then get busy and find out how to do it."

--- Theodore Roosevelt

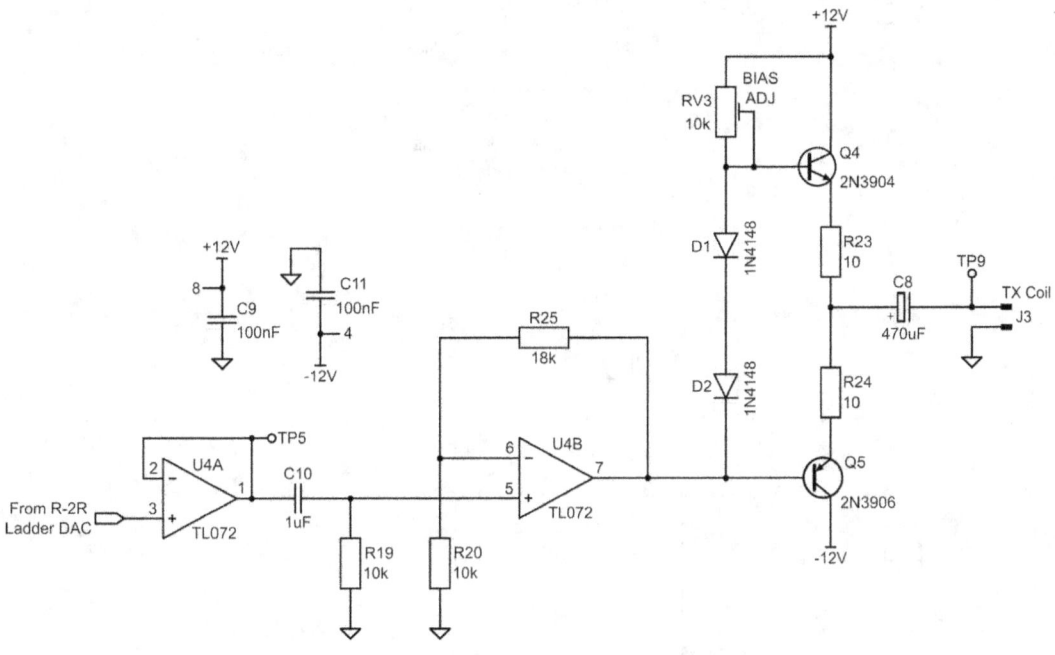

Fig. AB-1: TX Amplifier

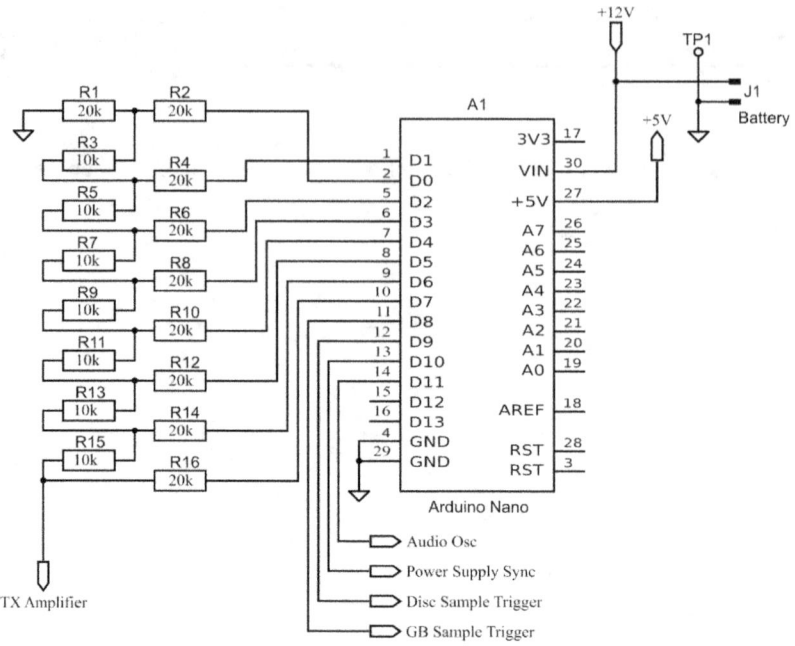

Fig. AB-2: Transmitter Oscillator

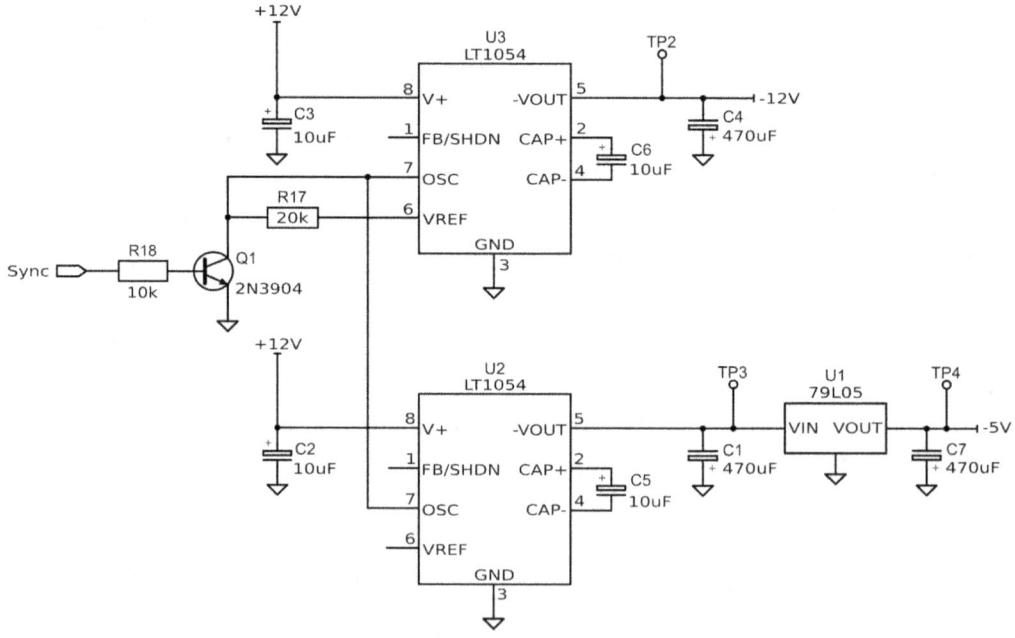

Fig. AB-3: Voltage Converters

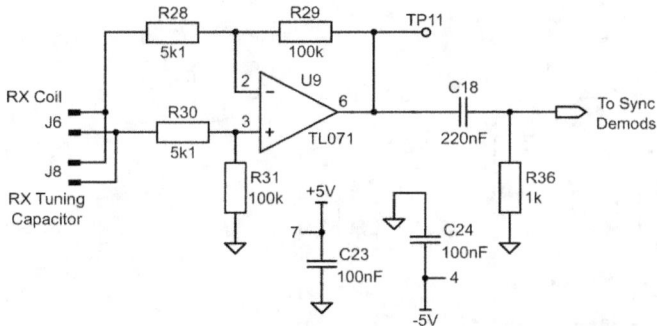

Fig. AB-4: Preamp

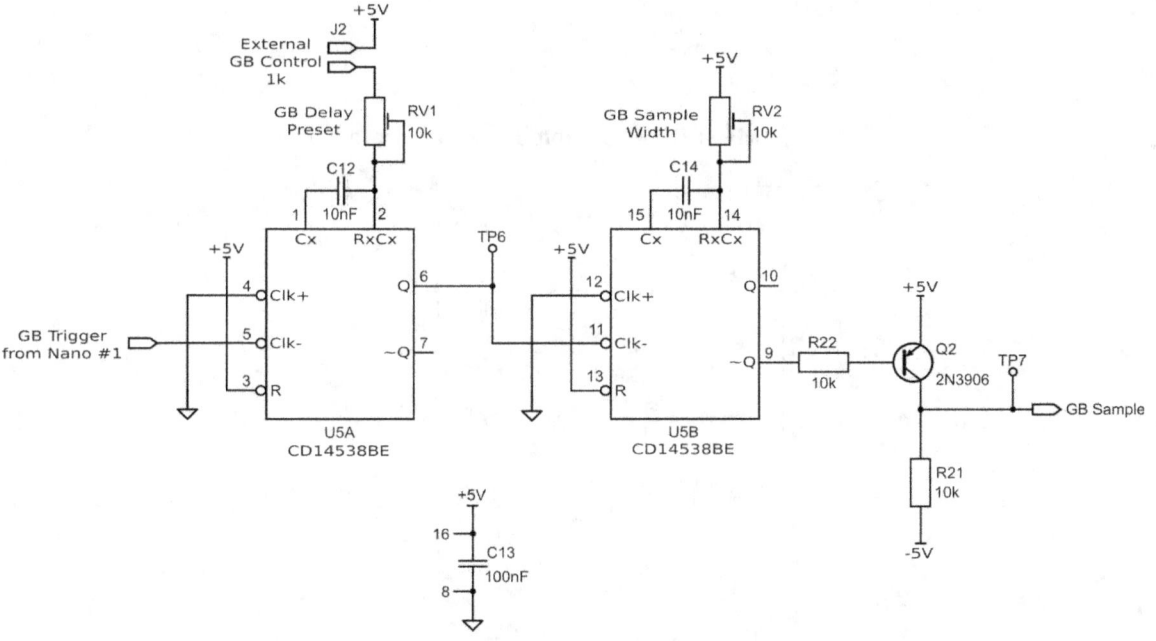

Fig. AB-5: GB Sample Pulse Generator

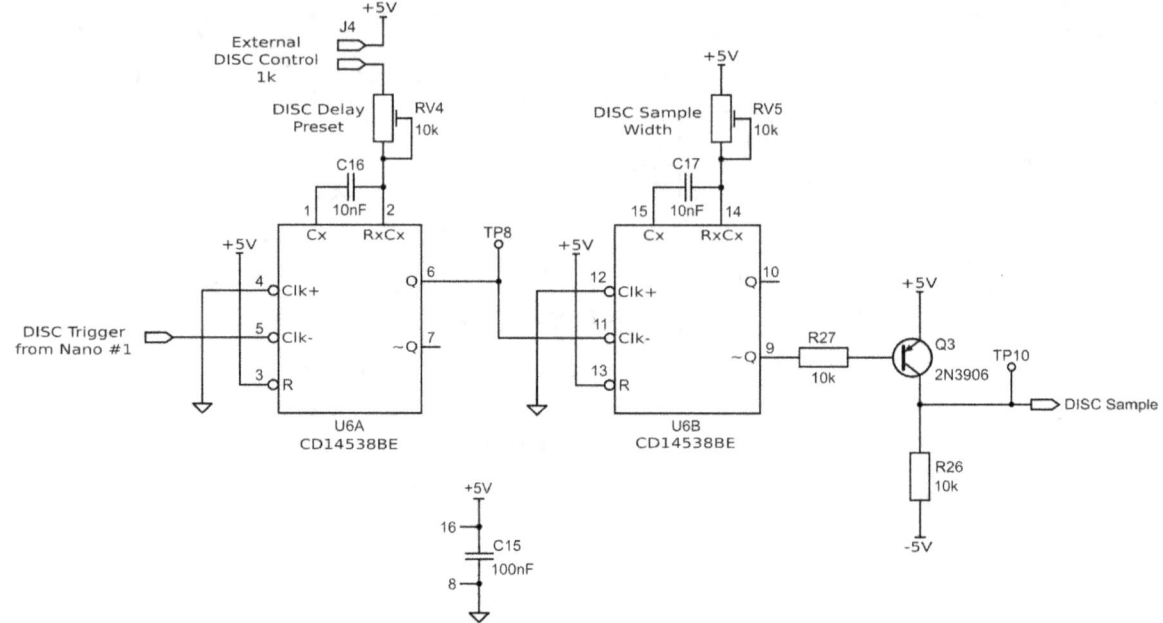

Fig. AB-6: DISC Sample Pulse Generator

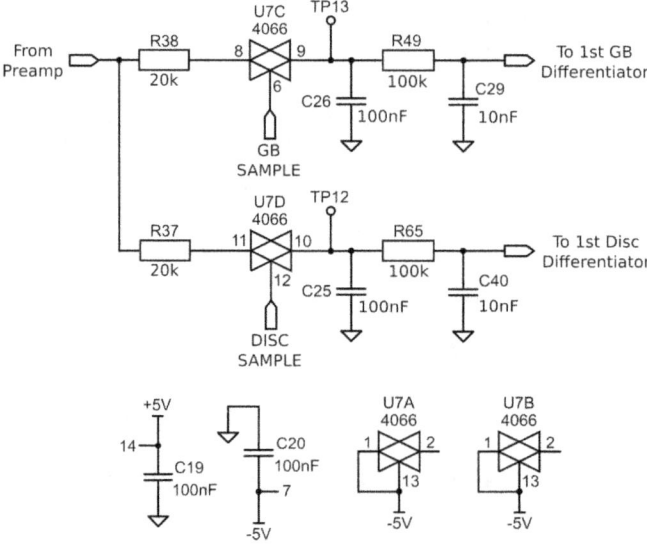

Fig. AB-7: Synchronous Demodulators

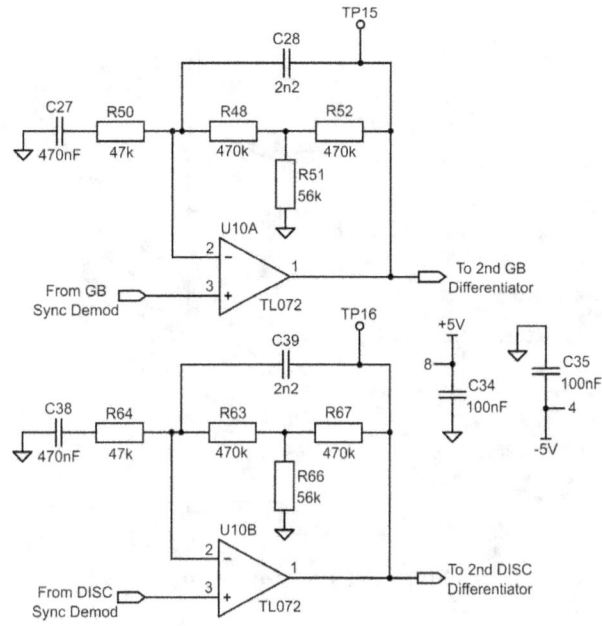

Fig. AB-8: Differentiators - First Stage (with T-network)

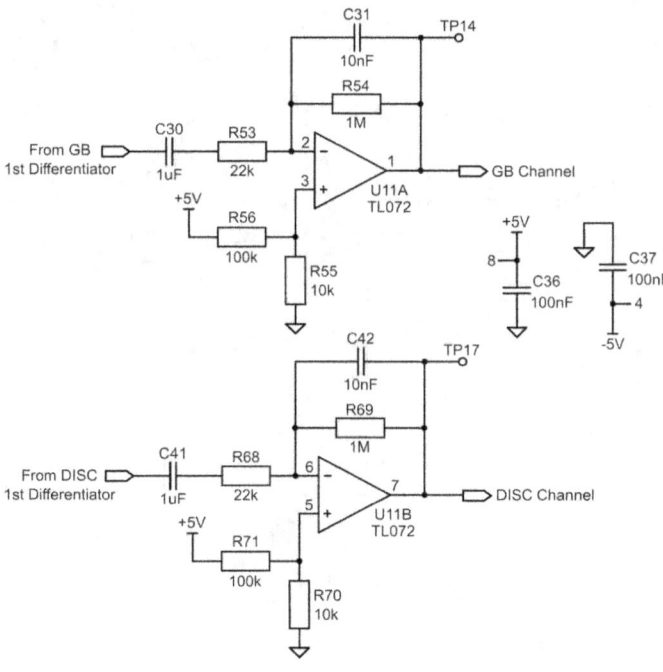

Fig. AB-9: Differentiators - Second Stage

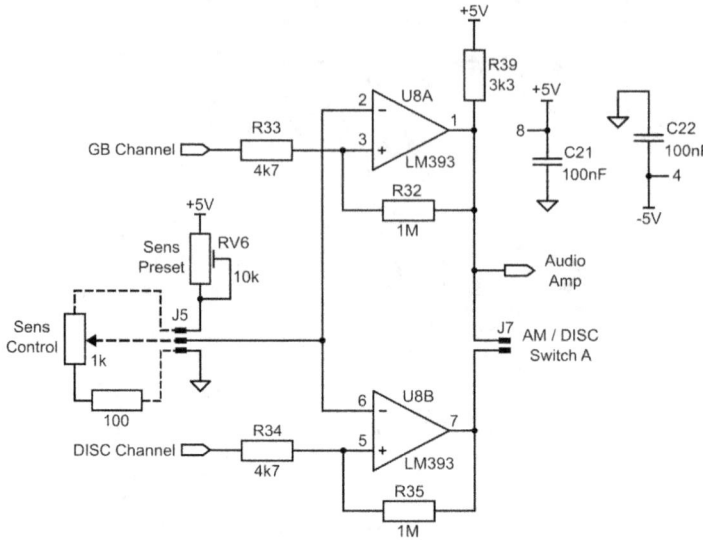

Fig. AB-10: Comparators

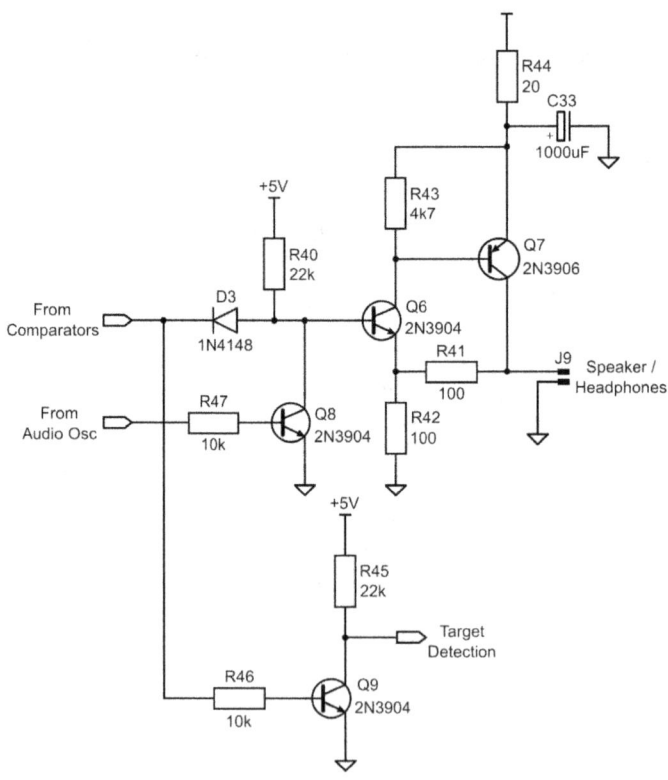

Fig. AB-11: Audio Amplifier

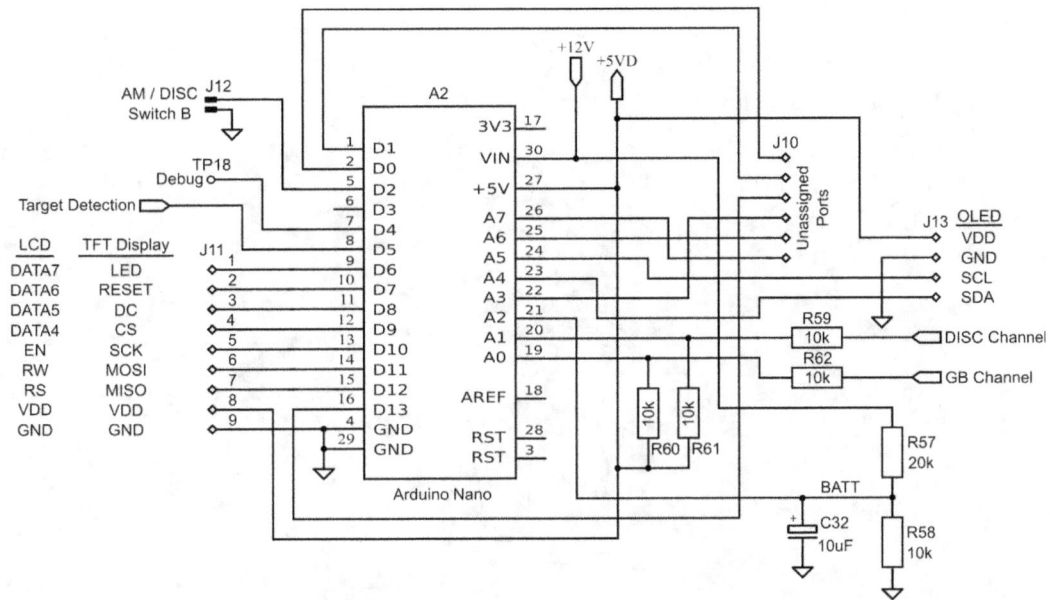

Fig. AB-12: Signal Processor and Display Controller

"Everything must be taken into account. If the fact will no fit the theory --- let the theory go."

--- Agatha Christie, The Mysterious Affair at Styles

LTspice can help you simulate electrical circuits before you build them, which means that you can make design changes without having to build any hardware. It can also be used to check your design after it has been built, and determine if it will work correctly in the real world once component tolerances are taken into account. One thing it cannot do is to design the circuit for you. For that you need to have at least some basic knowledge of electronics.

LTspice is a SPICE-based analog electronics circuit simulator from Analog Devices (originally by Linear Technology). SPICE stands for *Simulation Program with Integrated Circuit Emphasis*. Even though LTspice is freeware, there are no artificial restrictions imposed by the software.

In this chapter we will use LTspice to simulate various sections of the design to assist in better understanding how these circuits function. The simulation files will be available on the Geotech website.

R-2R Ladder DAC

In Chapter 2 we discussed the operation of an R-2R ladder DAC, and used a 4-bit example to examine how it works. Then it was explained how this circuit arrangement could be used to output a sinewave from an Arduino Nano that does not have a built-in D-to-A converter. You might be wondering why we then proposed the use of an 8-bit ladder DAC. Would a 4-bit DAC not suffice?

LTspice can be used to answer this question, as follows:

With reference to the circuit in Fig. AC-3, this is the same 4-bit R-2R ladder DAC as that shown in Fig. 2-1, but using "real" component values. This simulation is a little bit tricky to set up because we need to define 4 digital inputs such that the signal at node A is close to a sine wave. To make things a bit easier, I created a spreadsheet (using LibreOffice Calc) to help determine the logic inputs required at D0 to D3. This spreadsheet is shown in Fig. AC-1.

Since a 4-bit DAC has 16 possible logic combinations (0 to 15) then one time step will be approximately 5.21µs (22.5°) to generate one sine wave period at 12kHz. Hence column B increases downwards in steps of 22.5°. In column C, this is automatically converted to radians by dividing by 360 and multiplying by 2π. Column E then calculates the sine value for each step, multiplying this by 2.5 and adding 2.5 (as previously described in Chapter 2). This will produce a sine wave that has a peak-to-peak value of 5V centred around 2.5V. Then, in columns F to I, the logic states for each of the DAC inputs are determined. For example, for data point 5 the logic inputs are 1110, and the formulae for F9, G9, H9 and I9 are:

F9:   =IF(E9>=D3_value,1,0)

G9:   =IF(E9-F9*D3_value>=D2_value,1,0)

H9:   =IF(E9-F9*D3_value-G9*D2_value>=D1_value,1,0)

I9:   =IF(E9-F9*D3_value-G9*D2_value-H9*D1_value>=D0_value,1,0)

The numbers (D0_value to D3_value) contained in F2 to I2 are derived from Table 2-1, and represent the voltage associated with each DAC input.

	A	B	C	D	E	F	G	H	I	J
1	Data Points	Degrees	Radians	Time Steps	Sine Wave	D3	D2	D1	D0	Check Result
2						2.50	1.25	0.625	0.3125	
3										
4	0	0	0.000	0	2.500	1	0	0	0	2.500
5	1	22.5	0.393	1	3.457	1	0	1	1	3.438
6	2	45	0.785	2	4.268	1	1	0	1	4.063
7	3	67.5	1.178	3	4.810	1	1	1	1	4.688
8	4	90	1.571	4	5.000	1	1	1	1	4.688
9	5	112.5	1.963	5	4.810	1	1	1	1	4.688
10	6	135	2.356	6	4.268	1	1	0	1	4.063
11	7	157.5	2.749	7	3.457	1	0	1	1	3.438
12	8	180	3.142	8	2.500	1	0	0	0	2.500
13	9	202.5	3.534	9	1.543	0	1	0	0	1.250
14	10	225	3.927	10	0.732	0	0	1	0	0.625
15	11	247.5	4.320	11	0.190	0	0	0	0	0.000
16	12	270	4.712	12	0.000	0	0	0	0	0.000
17	13	292.5	5.105	13	0.190	0	0	0	0	0.000
18	14	315	5.498	14	0.732	0	0	1	0	0.625
19	15	337.5	5.890	15	1.543	0	1	0	0	1.250
20	16	360	6.283	16	2.500	1	0	0	0	2.500
21										

Fig. AC-1: Spreadsheet Used to Calculate 4-bit R-2R DAC Logic Inputs

Now you can already start to see the main drawback of using only 4 bits for the DAC. Column E shows the required sine wave values, whereas column J displays the actual output values. Comparing the two columns shows that there are some serious quantization errors, as the output resolution is quite large at 0.3125V.

The lack of resolution can be shown in a more graphical manner by plotting a line graph in LibreOffice Calc. This is shown in Fig. AC-2.

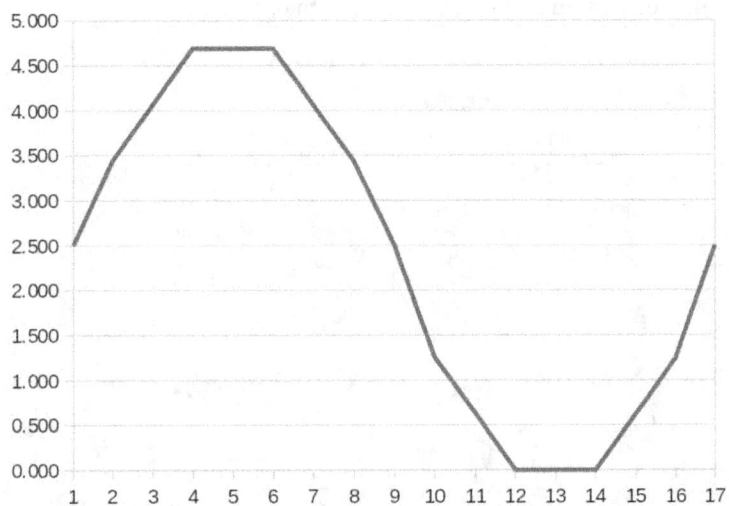

Fig. AC-2: Line Chart From Column J.

Now for an LTspice simulation:

With reference to Fig. AC-3, you can see that the digital inputs have been defined as piecewise linear (PWL) sources. The source data is contained in separate text files for each voltage source, and configured to repeat indefinitely. The text files were created by cutting and pasting from the spreadsheet, plus some editing to change the logic ones to "5" to represent the 5V logic level from the Arduino Nano.

LTspice was configured to run a transient analysis for 250µs to demonstrate that the sine wave repeats indefinitely The contents of the D0 to D3 text files are as follows:

D0.txt		D1.txt		D2.txt		D3.txt	
0.000E+00	0	0.000E+00	0	0.000E+00	0	0.000E+00	5
5.198E-06	0	5.198E-06	0	1.041E-05	0	4.687E-05	5
5.208E-06	5	5.208E-06	5	1.042E-05	5	4.688E-05	0
4.166E-05	5	1.041E-05	5	3.645E-05	5	8.332E-05	0
4.167E-05	0	1.042E-05	0	3.646E-05	0	8.333E-05	5
8.333E-05	0	1.562E-05	0	4.687E-05	0		
		1.563E-05	5	4.688E-05	5		
		3.124E-05	5	5.207E-05	5		
		3.125E-05	0	5.208E-05	0		
		3.645E-05	0	7.812E-05	0		
		3.646E-05	5	7.813E-05	5		
		4.166E-05	5	8.332E-05	5		
		4.167E-05	0	8.333E-05	0		
		5.207E-05	0				
		5.208E-05	5				
		5.728E-05	5				
		5.729E-05	0				
		7.291E-05	0				
		7.292E-05	5				
		8.332E-05	5				
		8.333E-05	0				

Note that the rise and fall times of the logic states have been defined as 10ns.

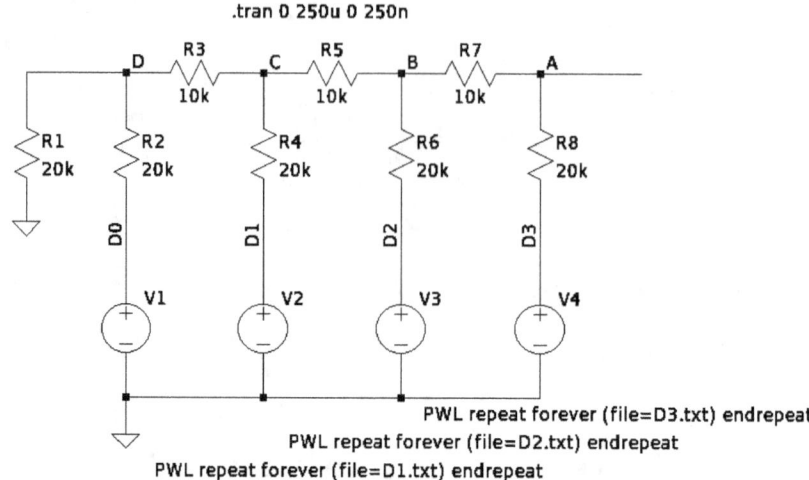

Fig: AC-3: 4-bit R-2R Ladder DAC

Comparing Fig. AC-4 to Fig. AC-2, there is an obvious difference between the two waveforms. This is because the spreadsheet chart erroneously does not take account of the fact that the Arduino Nano outputs switch between logic states in a discontinuous fashion, and do not simply linearly ramp between one state and the next. The LTspice simulation output demonstrates why is it important to have a larger number of bits than 4, and why we opted to use 8-bits. The 4-bit solution only provides 16 steps, compared to the 8-bit solution with 256.

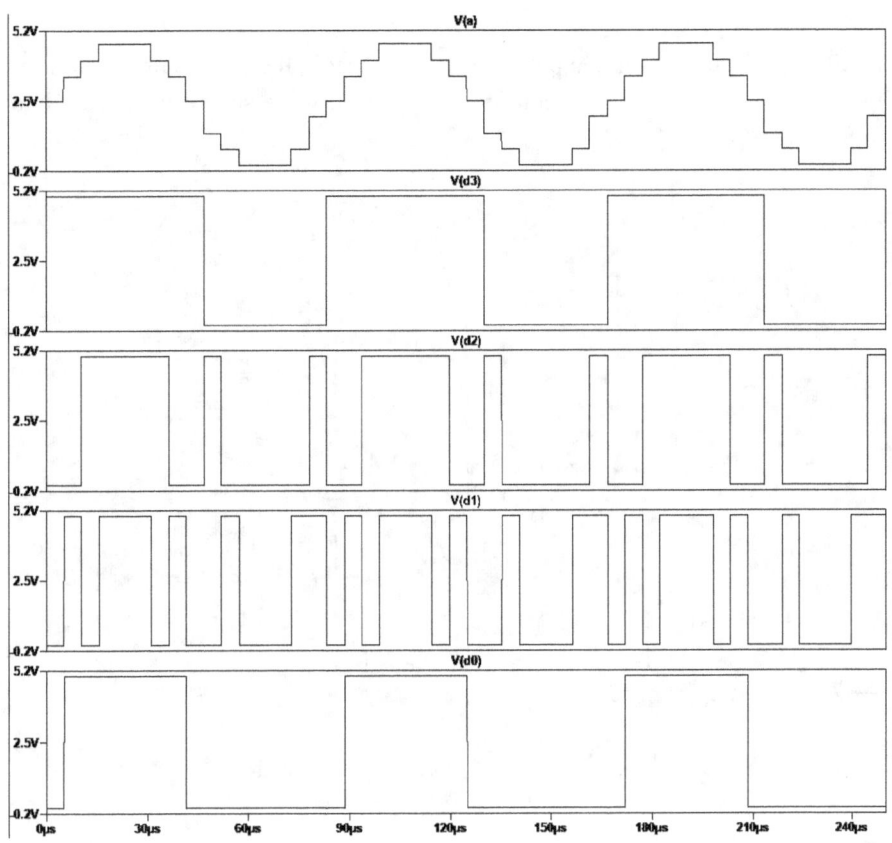

Fig. AC-4: 4-bit R-2R Ladder DAC Simulation Results

I will leave it to the reader as an exercise to duplicate the simulation using an 8-bit R-2R ladder DAC. Good luck!

TX Amplifier

In order to provide a powerful enough output to drive the TX coil, the DAC output needs to be amplified further. In Fig. AC-5 the DAC output is represented by a sine wave source that is fed into a push-pull output stage.

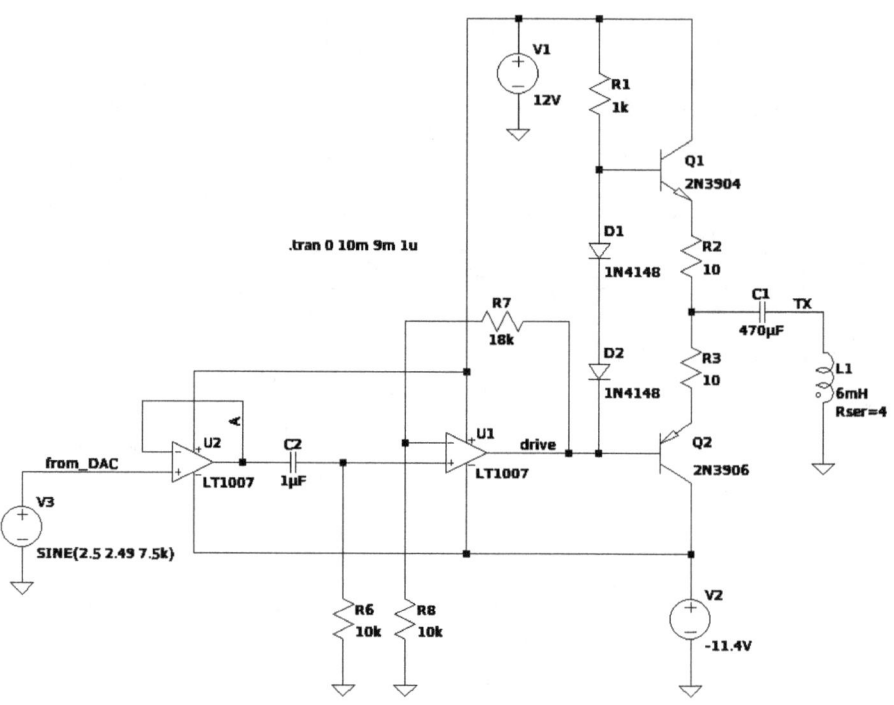

Fig. AC-5: TX Circuit

As explained in Chapter 3, the negative voltage rail is generated by a switched-capacitor voltage converter, and the output will always be slightly lower than the positive rail. This is why the negative voltage is set to -11.4V.

As you can see from Fig. AC-6 the DAC output has a peak-to-peak voltage of 14V, which compares well with the TX waveform shown in Fig. 2-6.

Since this VLF design uses a forced oscillator, you are free to choose whatever TX frequency you desire. In this particular simulation the TX frequency has been defined as 7.5kHz.

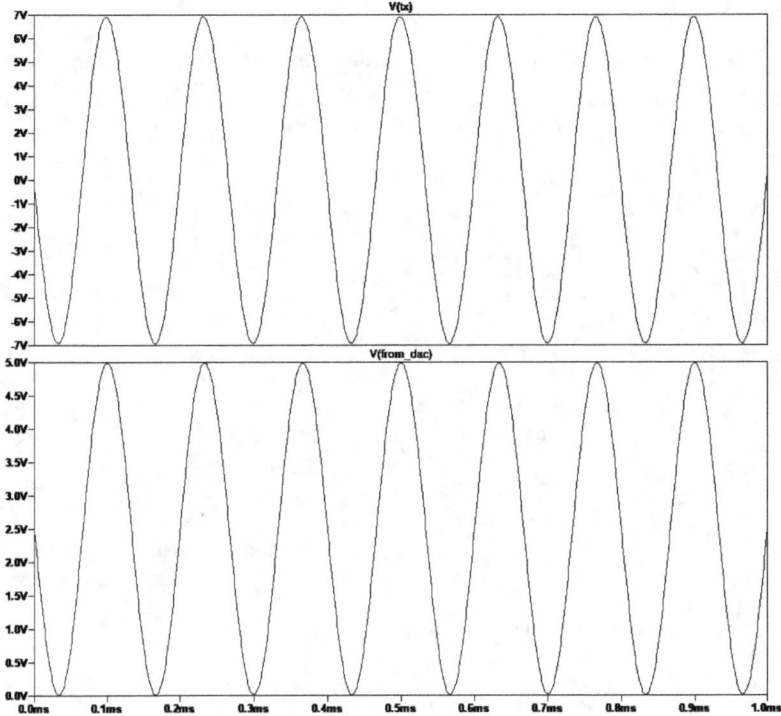

Fig. AC-6: TX and DAC Outputs

Complex Feedback Network

In Chapter 8 we explored the mysteries of using a complex network in the feedback path of an opamp, and the simulation below demonstrates that the results are identical to a standard differentiator circuit.

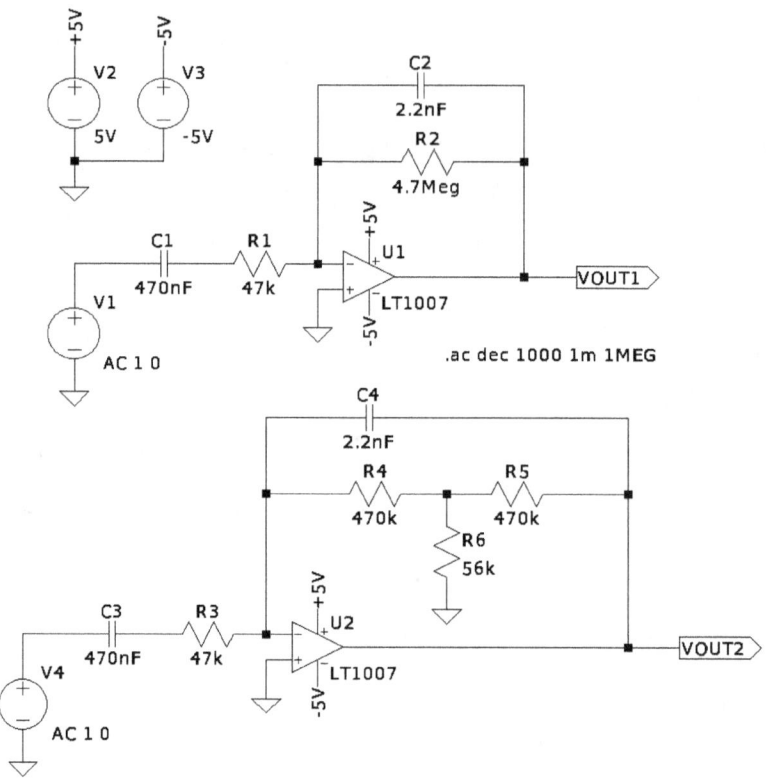

Fig. AC-7: Standard Differentiator Versus Complex Feedback Network

Note that the voltage sources (V1 and V4) are configured as AC sources. Since a frequency analysis is actually a small-signal analysis, and the simulation models are linearised around the operating point during circuit initialization, the input voltage can be set to 1 volt. This simplifies the process of comparing the output response to the input. Although the simulation was set up to use two separate AC sources in Fig. AC-7, we could in fact have used a single source to drive both circuits.

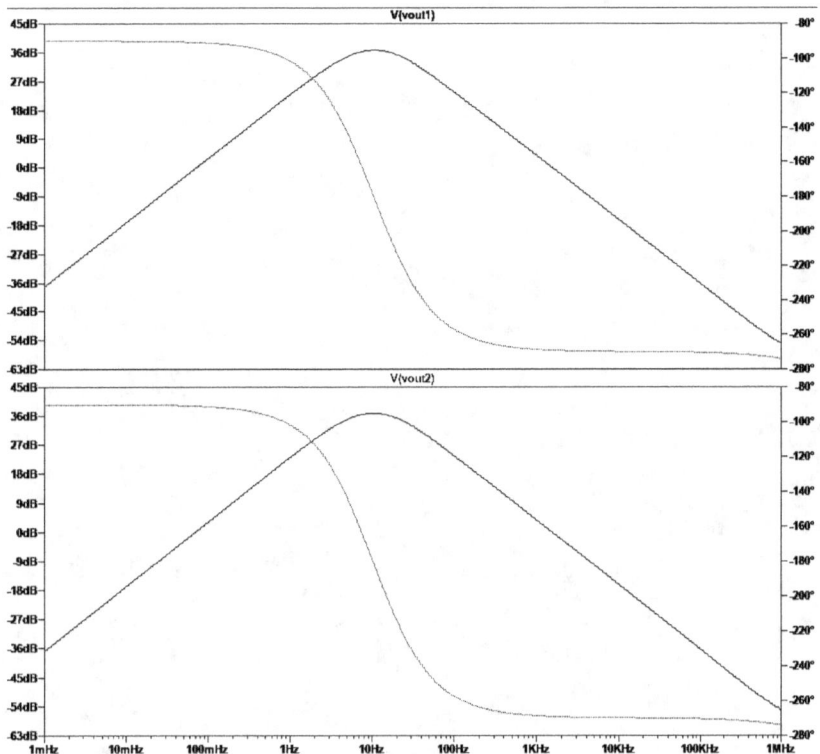

Fig. AC-8: Frequency Response for Both Circuits

Resources and References

"The most precious resource we have is time."

--- Steve Jobs

Geotech website: https://www.geotech1.com/forums – This site contains the single largest collection of metal detector information anywhere, with a large discussion forum on a variety of technical topics. A

useful air-cored coil calculator is available as a sticky post at the top of the Coils forum, and data files for the Arduino Nano VLF Metal Detector can be found in the *Technology* section under *Technical Books, Papers, and Literature*.

Inside the **METAL DETECTOR** (Second Edition)

George Overton & Carl Moreland – ISBN 9780985834210

The first in-depth book on metal detector technology since 1927.

It is aimed at anyone with an interest in metal detectors, who wants to understand in a clear and easy way how a metal detector works, irrespective of whether it is of the BFO, IB or PI variety.

All the supporting information and files for this project are available on the Geotech website.

The Voodoo Project – George Overton – ISBN 9798690296544

Pulse induction metal detectors are very sensitive to ferrous (iron) targets, and one that can ignore iron has been a sort of *holy grail* for metal detector developers for some time.

The Voodoo Project is a written record of the author's personal mission to design and develop a working pulse induction (PI) detector that is capable of good iron rejection.

Arduino Nano Pulse Induction Metal Detector Project

George Overton – ISBN 9798719427911

This book is intended for Arduino users who have already mastered the basics of programming, and for those who have at least an elementary knowledge of electronics. The project is for a pulse-induction (PI) metal detector with a professional level of performance.

KiCad – https://www.kicad.org/

KiCad is an open-source software suite for creating electronic circuit schematics, printed circuit boards (PCBs) and associated part descriptions. KiCad supports an integrated design workflow in which a schematic and corresponding PCB are designed together, as well as standalone workflows for special uses. KiCad also includes several utilities to help with circuit and PCB design, including a PCB calculator for determining electrical properties of circuit structures, a Gerber viewer for inspecting manufacturing files, and a 3D viewer for visualizing the finished PCB.

KiCad runs on all major operating systems and a wide range of computer hardware. It supports PCBs with up to 32 copper layers and is suitable for creating designs of all complexities. KiCad is developed by a volunteer team of software and electrical engineers around the world with a mission of creating free and open-source electronics design software suitable for professional designers. KiCad is available from the Linux software repository.

Schematics and PCB layout for the Arduino Nano VLF Metal Detector were created with this easy-to-learn software package. Gerber and NC drill files were then generated that can be sent to a PCB manufacturer.

Arduino - https://www.arduino.cc

Arduino is an open-source electronics platform based on easy-to-use hardware and software. It is intended for anyone making interactive projects.

Everything you wish to know about Arduino can be found at the above URL, including hardware, software, and documentation.

Linux Mint - https://linuxmint.com

The purpose of Linux Mint is to produce a modern, elegant and comfortable operating system which is both powerful and easy to use. It is one of the most popular desktop Linux distributions and used by millions of people.

Here's why you should to use Linux Mint:

- It works out of the box, with full multimedia support and is extremely easy to use.
- It's both free of cost and open source.
- It's community-driven. Users are encouraged to send feedback to the project so that their ideas can be used to improve Linux Mint.
- Based on Debian and Ubuntu, it provides about 30,000 packages and one of the best software managers.
- It's safe and reliable. Thanks to a conservative approach to software updates, a unique Update Manager and the robustness of its Linux architecture, Linux Mint requires very little maintenance (no regressions, no antivirus, no anti-spyware...etc.).

Linux Mint was the only operating system used during the development of the Arduino Nano PI detector project, and also in writing this book.

LibreOffice - https://www.libreoffice.org

LibreOffice is a powerful and free office suite, a successor to OpenOffice(.org), used by millions of people around the world. Its clean interface and feature-rich tools help you unleash your creativity and enhance your productivity. LibreOffice includes several applications that make it the most versatile Free and Open Source office suite on the market: Writer (word processing), Calc (spreadsheets), Impress (presentations), Draw (vector graphics and flowcharts), Base (databases), and Math (formula editing).

This book was created using LibreOffice Writer, which is a modern, full-featured word processing and desktop publishing tool.

Inkscape - https://inkscape.org

Inkscape is a Free and open source vector graphics editor for GNU/Linux, Windows and MacOS X. It offers a rich set of features and is widely used for both artistic and technical illustrations such as cartoons, clip art, logos, typography, diagramming and flowcharting. It uses vector graphics to allow for sharp printouts and renderings at unlimited resolution and is not bound to a fixed number of pixels like raster graphics. Inkscape uses the standardized SVG file format as its main format, which is supported by many other applications including web browsers.

It can import and export various file formats, including SVG, AI, EPS, PDF, PS and PNG. It has a comprehensive feature set, a simple interface, multi-lingual support and is designed to be extensible; users can customize Inkscape's functionality with add-ons.

GIMP - https://www.gimp.org

GIMP is an acronym for GNU Image Manipulation Program. It is a freely distributed program for such tasks as photo retouching, image composition and image authoring.

FreeCAD - https://www.freecadweb.org

FreeCAD is an open-source parametric 3D modeller made primarily to design real-life objects of any size. Parametric modelling allows you to easily modify your design by going back into your model history and changing its parameters.

It is also multiplatform (Windows, Mac and Linux) and highly customizable and extensible. It reads and writes to many open file formats such as STEP, IGES, STL, SVG, DXF, OBJ, IFC, DAE and many others.

FreeCAD was used in this project to create some 3D printable fixtures.

Repetier-Host - https://www.repetier.com

The Repetier-Host program is used in conjunction with built-in slicing software to prepare 3D models for printing. You can either draw a 3D model yourself (e.g. using FreeCAD, and exporting an STL file) or download an STL file from the web. Repetier-Host will load the model into a graphical work space in which you can move and rotate the piece if needed. After slicing the model you can generate a *gcode* (and/or *gco*) file to send to a 3D printer.

Geeetech - https://www.geeetech.com

A Geeetech I3 pro B 3D printer was used to print a few 3D fixtures for the Arduino Nano VLF Metal Detector. This is a 3D printer designed and manufactured by Shenzhen Getech Co., Ltd. It supports 5 different types of filament: ABS, PLA, Wood-Polymer, Nylon and flexible PLA.

Amazon KDP - https://kdp.amazon.com

KDP stands for Kindle Direct Publishing, and allows people to self-publish eBooks and paperbacks without involving a traditional publishing house. This book has been published and distributed exclusively by Amazon, and is only available in paperback format.